Experimental Design and Verification of a Centralized Controller for Irrigation Canals

Satisfaction lies in the effort, not in the attainment, full effort is full victory
Mahatma Gandhi

To my parents, Antonio, Angeles and Tere
My strength and my life

Series Editor
Gilles Pijaudier-Cabot

Experimental Design and Verification of a Centralized Controller for Irrigation Canals

Enrique Bonet Gil

ISTE PRESS

ELSEVIER

First published 2018 in Great Britain and the United States by ISTE Press Ltd and Elsevier Ltd

ISTE Press Ltd
27-37 St George's Road
London SW19 4EU
UK

www.iste.co.uk

Elsevier Ltd
The Boulevard, Langford Lane
Kidlington, Oxford, OX5 1GB
UK

www.elsevier.com

British Library Cataloguing-in-Publication Data
A CIP record for this book is available from the British Library
Library of Congress Cataloging in Publication Data
A catalog record for this book is available from the Library of Congress
ISBN 978-1-78548-307-3

Contents

Preface

This book aims to develop a predictive controller for irrigation canals to improve the management of water resources.

Water is necessary for life and it is a scarce good that we need for drinking, for agriculture, etc. At the same time, it can constitute a serious threat in particular areas due to the difficulty of growing foods as a result of the increase of prolonged droughts.

Agriculture is an important part of the food chain and the water resources for agriculture are crucial. The problem is that water transport systems present low efficiencies in practice. The agriculture yield has to be optimized, because the goal of an operational water manager is to deliver the water to the irrigation sites accurately and efficiently. To improve the efficiency of the water transport systems, it is necessary to invest in automating the operation of irrigation canals.

The management of a canal starts from setting the demand delivery accurately while taking into account the crops' necessities during the irrigation cycle and establishing the gate trajectories for controlling the canal at each time step. In an ideal world, the system would be controlled; but someone could introduce a disturbance in the canal which could deviate the real canal state from the desired canal state. In this circumstance, it would be necessary to have a feedback controller which could help the watermaster restore the desired canal state. In order to fulfill this objective, we define an overall control diagram scheme in Chapter 4 which splits the management of the canal control into different blocks. Each of these blocks is represented by a particular algorithm. The algorithms developed and tested in this book are the CSI and GoRoSoBo algorithms.

The CSI algorithm defines a powerful tool in the management of a canal. The watermaster establishes the gates' positions and fixes the desired water level at checkpoints to fulfill a scheduled demand. In this regard, when someone introduces a disturbance into the canal, they perturb the water level at checkpoints, so the scheduled deliveries cannot be fulfilled by the watermaster. In such a case, the water-level measurements at these checkpoints could be sent to the CSI algorithms which calculate the real extracted flow and the current canal state along the canal, that is, the water level and the velocity in all cross-sections of the canal. This task is performed by the CSI algorithm which has been designed in this book and tested in numerous numerical examples (see Chapter 6) and experimentally in the laboratory canal of the Technical University of Catalonia (see Chapter 7).

The GoRoSoBo algorithm is the essential tool in the management of a canal, that is, a control algorithm operating in real time. The GoRoSoBo algorithm (Gómez, Rodellar, Soler and Bonet) is a feedback control algorithm that calculates the optimum gates trajectories for a predictive horizon, taking into account the current canal state and the real extracted flow obtained by CSI as well as the scheduled demands and the previous gate trajectories. GoRoSoBo has been developed in this book and tested in several numerical examples (see Chapter 9) as well as with the test cases proposed by the ASCE Task Committee on Canal Automation Algorithms (see Chapter 10). In this regard, we propose a centralized control performance to manage the canal control.

In addition to these two main contributions, many other smaller developments, minor results and practical recommendations for irrigation canal automation are presented throughout this book.

Enrique BONET GIL
May 2018

Acknowledgments

I could never understand why there are so many thank yous written in the acknowledgments of a book; now that I am publishing my own I can begin to understand. Too much time is invested not to ask for favors and not to receive any help.

I thank my thesis tutors Manuel Gómez and Joan Soler for the time they dedicated, their patience and their support along this long journey that we have traveled together. I appreciate you enormously, thank you very much.

Ivan, I will never forget you, my friend! Thank you Cesca for your time!

Thanks Javi for the help programming the displayer in Java, you were a great teacher and an even better person!

Thank you to my uncles, Emilio and Olga, who I have always been able to count on and who I will never forget; I love you. To Ana and my cousin Emiliet, I am very proud of him and I hope that he keeps fighting as he constantly does.

Thanks to Isidro and Felicidad, for your constant support, to Javi and Juani and to the kids.

Well, it is time to thank people who are an important part of my life and it is difficult not to get excited. Tere, for being my partner, friend and always encouraging me, thanks for all of these years that you have been a part of me; together there is nothing that is impossible.

To my parents, my life, thank you for giving me your principles, defects, virtues and infirmities, I like it all. I love you.

Thank you grandfathers, you always loved me very much, I hope to see you someday; I never said goodbye. Someone once said that a person never dies until the last person who loved them lives; you will always live with me.

Hasta la victoria siempre!

Nomenclature

α:	Index defined in the Active Set method
Δ:	Divergence of a vector field's source or sink
λ_k:	Lagrange multipliers of active constraints of the quadratic problem at iteration k
λ:	Lagrange multipliers vector at iteration k
Λ:	Diagonal matrix with the eigenvalues of the G matrix
ν:	Marquardt coefficient
γ:	Prediction horizon in CSI
φ:	Prediction horizon in GoRoSoBo
ψ:	Parameter to denote variables of flow description such as depth and velocity
A:	Matrix of dimension $(2\times n_S) \times (2\times n_S)$, obtained defining the HIM(Q)
A(y):	Area of the wet section which depends on the water level "y"
a_c:	Gate width
B:	Vector of dimension $(2\times n_S)$, $b[x^{k+1}, q(K)]$, obtained defining the HIM(Q)
b_j:	j-gate width
c(y):	Water wave celerity which is dependent on the water level "y"

C:	Matrix of dimension $n_Y \times n_X$ called "discrete observer matrix" by Malaterre
C_c:	Contraction coefficient of the gate
C_d:	Discharge coefficient of the gate
c_T:	Local losses coefficient in the canal
C_w:	Discharge coefficient of a weir
d_j:	Drop at j-gate
d_0U_{max}:	Maximum gate movement allowed between two consecutive operation periods
dU_{max}:	Maximum gate movement allowed between two consecutive operation periods of different prediction horizons
ΔU:	Perturbed gate trajectory vector
ΔQ_b:	Perturbed input/output flow vector
Δt:	Numerical discretization time according to CFL conditions
ΔT:	Operation period defined by the watermaster
ΔY:	Water level perturbation
$\Delta\Delta U$:	Auxiliary solution-vector (gate) of the quadratic problem
Δx:	Numerical discretization space cell length
ΔX:	Perturbed state vector obtained when a perturbation is introduced into the system
$error_{k_1}^{k_F}$:	Error between the computed and desired water level at a checkpoint
f(k):	Input function at time step k
g:	Gravity
G_{it}:	Performance index Hessian approximation at iteration it
$g_i(X)$:	Inequality constraints
$G(Q_b)$:	Set of feasible directions for the extracted flow vector
G:	Matrix obtained multiplying the transposed "J" matrix by the "J" matrix

G(U):	Set of feasible directions for the gate trajectories vector
$\mathbf{g}_{it}$:	Auxiliary Jacobian of the Lagrange's function at iteration *it*
H_{it}:	Auxiliary Hessian approximation of the Lagrange's function at iteration it
HIM(Q_b):	Hydraulic influence matrix (derivative parameter Q_b)
HIM(U):	Hydraulic influence matrix (derivative parameter U)
H_{up}:	Constant upstream water level at upstream boundary condition (canal header)
I:	Identity matrix
$[I_M(q_b)]_j$:	Hydraulic influence on the state vector of a certain pump station "j"
$[I_M(u)]_j$:	Hydraulic influence on the state vector of a certain gate "j"
J(Q_b):	Performance criterion (input/output flow trajectory) (objective function)
J(U):	Performance criterion (gate trajectory) (objective function)
J:	Matrix obtained deriving the output water level vector by a variable
k_F:	Number of sections in which the prediction horizon is divided depending on the CFL condition/time frequency of measuring data
K_F:	Number of operation periods defined by the watermaster in which the simulating horizon is divided
K_p:	Proportional gain, a tuning PID parameter
K_i:	Integral gain, a tuning PID parameter
K_d:	Derivative gain, a tuning PID parameter
$k_o(k)$:	Orifice coefficient depending on its overture of the gravity i-offtake at time instant k
L_n:	Water level from sensor called Ln
L:	Lower triangular matrix set to Cholesky factorization
L':	Matrix with the direct influence from inlet–outlet flow discharge over the canal

$l_j(X)$:	Equality constraints
$\mathcal{L}$:	Lagrangian function
M:	Matrix obtained from the Saint-Venant equation to represent the influence between parameters
N:	Matrix obtained from the Saint-Venant equation to represent the influence between parameters
n:	Manning coefficient
n_c:	Number of checkpoints
n_g:	Number of gates
n_{HX}:	Number of rows of the hydraulic influence on the state vector of a certain gate "j"
n_{HY}:	Number of columns of the hydraulic influence on the state vector of a certain gate "j"
n_p:	Number of pump stations
n_q:	Dimension of inlet–outlet flow vector ($n_Q=n_P\times K_F$)
n_s:	Number of sections in which the canal is discretized
n_u:	Dimension of control trajectory vector ($n_U=n_g\times K_F$)
n_x:	Dimension of prediction vector ($n_X=(2\times n_S)\times k_F$)
n_y:	Dimension of prediction output vector ($n_Y=k_F\times n_C$)
$o_{k_1}^{k_F}$:	Error between the computed and desired flow rate at a checkpoint
P(y):	Wet perimeter which is function of the water level "y"
Q_b:	Input/output flow vector
Q':	Weighting matrix
Q_b^*:	Solution of the input/output flow vector
$\bar{Q}_b$:	Desired input/output flow vector
$q_b(K)$:	Inlet–outlet flow at the control structure during the operation period K
Q_j:	Undershot j-gate flow

Q_{wn}:	Discharge over the weir n
$q_{offtake}$:	Discharge through the orifice offtake
q_s:	Discharge through the lateral spillway
R_H:	The hydraulic radius is a measure of a channel flow efficiency
$r_k(u)$:	Constraint function "k" applied to the gate trajectory
$r_{k_1}^{k_F}$:	Residual vector between the desired water level and the computed water level
S:	Matrix obtained from the Saint-Venant equation to represent the influence between parameters
S(y):	Horizontal surface of the reception area in the checkpoint
S_0:	Bottom slope
S_f:	Friction slope
$s_{k_1}^{k_F}$:	Error between the computed and the minimum operating cost gate trajectory at a checkpoint
s(k):	Input data vector at time step k
T^K:	Operation instant when the inlet–outlet flow is changed
T(y):	The maximum width dependent on the water level "y"
$u_{ij}(K)$:	Gate-opening of the gate i during the operation period K at the prediction horizon j
U:	Gate trajectory vector
U^*:	Solution of the gate trajectory vector
$\bar{U}$:	Desired gate trajectory vector
U_0:	Vector with the gate positions at the first regulation period for the previous predictive horizon
U_{max}:	Maximum gate position allowed
U_{min}:	Minimum gate position allowed
$U_{ij}(k)$:	Gate-opening of the gate i at the predictive horizon j for the regulation period k

$v_i(k)$:	Mean velocity at time instant k at canal section i
V:	Orthonormal matrix defined by the eigenvectors of G
$W(Q_b)$:	Hessian matrix obtained deriving the variable Q_b (pump flow vector)
W(U):	Hessian matrix obtained deriving the variable U (gate trajectories vector)
W_n:	Weir number "n"
x(k):	State vector at time instant k
$\mathbf{X}_1^{k_F}$:	Prediction vector from 1 to k_F
y(k):	Subset of water depths of the state vector at time instant k at checkpoints
$\mathbf{Y}_1^{k_F}$:	Prediction output vector from 1 to k_F
$y_M(k)$:	Subset of measurement water levels at checkpoints at time instant k
$\mathbf{Y}_{M_1}^{k_F}$:	Measurement output vector
y_{dw}:	Downstream j-gate water depth
$y_i(k)$:	Water depth at time instant k at canal section i
y_o:	Height of the center of the orifice from bottom of the gravity offtake
y_s:	Downstream water level of the reservoir
y_{up}:	Upstream j-gate water depth
z:	Variable used to interpolate values with the Lagrange factors

Abbreviations

ASCE:	American Society of Civil Engineers
AEMET:	Agencia Estatal de Meteorología
AR4:	Rossby Center regional atmospheric model
BIVAL:	Constant downstream volume system
Canal PAC-UPC:	Canal de Prueba de Algoritmos de Control – Universitat Politècnica de Catalunya
CACG:	Compagnie d'Aménagement des Coteaux de Gascogne
CARA:	Compagnie d'Aménagement Rural d'Aquitaine
CARDD:	Canal Automation for Rapid Demand Deliveries
CEMAGREF:	Centre National du Machinisme Agricole, du Génie Rural, des Eaux et des Forêts
CFL:	Courant Friedrichs Levi
CLIS:	Constant Level control method based on the Inverse Solution of the Saint-Venant equations
CNABRL:	Compagnie Nationale d'Aménagement du Bas-Rhône Languedoc
CPU:	Central Processing Unit
CSI:	Canal Survey Information
ELFLO:	Electronic Filter and Level Offset

FAO:	Food and Agriculture Organization of the United Nations
HEC-RAS:	Hydrologic Engineering Center – River Analysis System
HIM:	Hydraulic Influence Matrix
HIRLAM:	High Resolution Limited Area Model
GN:	Gauss Newton
GoRoSo:	Gómez, Rodellar and Soler (feedforward algorithm)
GoRoSoBo:	Gómez, Rodellar, Soler and Bonet (feedback algorithm)
GPC:	Generalized Predictive Control method
IAE:	Integral of Absolute Magnitude of Error
IAQ:	Integrated Absolute Discharge Change
ICT:	Information and Communications Technology
ID:	Integrator Delay
INE:	Instituto Nacional de Estadística
LQ:	Linear Quadratic
LQG:	Linear Quadratic Gaussian
LQR:	Linear Quadratic Regulator
LM:	Levenberg–Marquardt
LN:	Lagrange Newton
MAE:	Maximum Absolute Error
MCG:	Modelos de Circulación General
MCR:	Modelos Climáticos Regionales
MIMO:	Multiple-Input Multiple-Output
MISO:	Multiple-Input Single-Output
MMA/MIMAM:	Ministerio de Medio Ambiente
MPC:	Model Predictive Control
PC:	Predictive Control

PILOTE:	Preissmann Implicit scheme, Linear Optimal control, Tracking of variables and Estimation of perturbations
PI:	Proportional Integral
PID:	Proportional Integral Derivative
PIR:	Proportional Integral Retarded
P.K.:	Distance along an axis from the origin point
PNACC:	Plan Nacional de Adaptación al Cambio Climático
PRECOM:	Multivariable Predictive Controller
PREDAPTROL:	Predictive and Adaptive control
PREDTROLLER:	Predictive controller
QP:	Quadratic Programming
SACMAN:	Software for Automated Canal Management
SCADA:	Supervisory Control And Data Acquisition
QP:	Quadratic Programming
SIC:	Simulation of Irrigation Canals
SIMO:	Single-Input Multiple-Output
SISO:	Single-Input Single-Output
STE:	Steady-state Error
UMA:	PID developed by UMA engineering
UNDESA:	United Nations Department of Economic and Social Affairs

Introduction

An irrigation canal is a hydraulic system whose main objective is to convey water from a source (dam and river) to different users. Such systems can be very large (several tens or hundreds of kilometers); they are characterized by time delays and nonlinear dynamics, strong unknown perturbations and interactions among subsystems. Varying operational objectives are assigned to their managers. The main general objective is to provide water to the different users at the right moment and in the right quantity, reducing losses as much as possible in order to guarantee the safety of the infrastructure. In particular, a major concern is to prevent the canals from overtopping, and also from having water levels inside the pools below the supply depths of the gravity offtakes.

One of the most important problems for an on-demand irrigation system designer is the calculation of the discharges flowing into the network. Such discharges strongly vary over time depending on the cropping pattern, meteorological conditions, on-farm irrigation efficiency and farmers' behavior. The amount of water to be supplied during an irrigation event, referred to as the target or required depth of application, is a major design consideration.

The unknown changes in water demand are quite normal in a canal during an irrigation cycle. The result is that the real extracted flow is unknown to the watermaster. In this case, it is quite difficult to recompute a new gate trajectory to satisfy the scheduled deliveries when the watermaster does not know the real extracted flow and the current canal state. For this reason, the motivation of this book is centered on two main objectives. On the one hand, the CSI algorithm is developed to establish the disturbances and the current canal state, that is, an algorithm that calculates the velocity and water level in each cross-section of the canal from water depth levels at certain canal sections during a past time horizon. On the other hand, a predictive control algorithm (GoRoSoBo algorithm) is used to establish the gate trajectories in real time from the current canal state. In this way, the watermaster could fulfill the real water demands of the canal.

The book is divided into 10 chapters:

In Chapter 1, we first talk about estimations about agricultural water consumption, and then define the main problems generated by low efficiency water use. To fix the problems related with low efficiency, it is necessary to point out the main issues: water distribution infrastructure (the transport and the techniques of application of water parcels), water prices and the productive structure, that is, the economic dimensions of irrigated farms. To finalize the chapter, we define the concept of diagnosis and prognosis in canal control, check the efficiency of water use obtained by control algorithms in the diagnosis, and propose mechanisms to improve the results of these algorithms in the prognosis.

In Chapter 2, we provide a literature review of canal control algorithms. We define important control concepts which help us to introduce a classification of control algorithms in irrigation canals. Each of these algorithms has been characterized according to several criteria. We follow the criteria established by Malaterre [MAL 98]. We classify these algorithms by the considered variables (controlled, measured and control action variables), the logic of control (type and direction), the design technique and the field implementation.

In Chapter 3, we present a general introduction to the main characteristics of control algorithms and define their main elements. We give our particular point of view on predictive control and introduce our computer model and optimization problem, as well as their main elements.

In Chapter 4, the modernization of irrigation projects is achieved via a general improvement in water transport and distribution networks and in field application techniques with a view to enhancing the efficiency of water use, which is extremely low. Such an overall improvement in efficiency can occur at three different levels: transport through the main canal, distribution through the secondary network and field application. In our opinion, improving water use at these three levels must be tackled at each level individually and together. In line with a predictive control theory [MAR 96], we define in this chapter the different calculation steps that should be included in our overall canal control irrigation process, such as the crop needs and outlet hydrographs, parameter identification, offline reference trajectories, canal current state and online predictive control.

In Chapter 5, we introduce the CSI (Canal Survey Information) algorithm which sets the real extracted flow and the hydrodynamic state in a canal at any moment from water depth levels at certain canal sections. We develop this algorithm for several reasons: it is a useful tool to get information about disturbances and it is a useful supplementary tool for a feedback controller since this algorithm establishes the past and current hydrodynamic state of a canal, that is, the water level and flow rate at every cross-section of the canal from the past to the present time. To develop this algorithm, we also had to develop the hydraulic influence matrix, which establishes the influence of flow changes (for instance, one pump running) on the canal state vector.

In Chapter 6, the CSI algorithm is tested through some numerical examples. We define a main canal system consisting of two canal pools, and we test the CSI algorithm in this canal in three different tests. We also consider a sensitivity analysis which we did to evaluate the CSI algorithm when obtaining errors in water-level measurements from the canal sensors.

In Chapter 7, we test the CSI algorithm in a laboratory canal. The laboratory canal used is the Canal PAC-UPC, which is designed to develop basic and applied research in irrigation canal control areas and in all subjacent areas like canal instrumentation, canal modeling and water measurements. The CSI algorithm is tested in the canal and we present the results. We introduce several tests to study the sensitivity of the algorithm versus changes in physical parameters such as the Manning coefficient.

In Chapter 8, we introduce and develop the theory of the proposed feedback algorithm, GoRoSoBo (Gómez, Rodellar, Soler and Bonet). GoRoSoBo is a predictive control algorithm that maintains a target level at several sections of the canal, so that the flow through the offtakes is according to scheduled demands, in case of unknown flow changes in the canal. Any water level deviation from the desired set point is sent to the CSI algorithm which computes the real extracted flow and the current canal state, and this data is sent to the GoRoSoBo algorithm in order to produce a corrective action (new gate trajectories). To develop this algorithm, we used the hydraulic influence matrix, which establishes the influence of a gate movement on the canal state vector, introduced by Soler [SOL 03]. Furthermore, we define the algorithms developed in each block of our overall control scheme introduced in Chapter 4.

In Chapter 9, the GoRoSoBo algorithm is tested by several numerical examples. We define a main canal system consisting of one canal pool, and we test the GoRoSoBo algorithm in this canal in eight different tests.

In Chapter 10, the GoRoSoBo algorithm is applied to the "test case" examples defined by the ASCE Task Committee on Canal Automation Algorithms. We analyze the results obtained with GoRoSoBo and compute the performance indexes proposed by the ASCE. We compare the performance indicators obtained with GoRoSoBo with the performance indicators estimated by other centralized predictive controllers such as CLIS [LIU 98] and Pilote [MAL 95].

We end this book with a general conclusion and discussion about future work.

1

Introduction and General Statements

The world population is predicted to grow from 6.9 billion in 2010 to 8.3 billion by 2030 and 9.1 billion by 2050 (UNDESA, 2009), and food demand is predicted to increase by 50% by 2030 and by 70% by 2050 [BRU 09].

The most recent estimates for irrigated agriculture suggest an increase from 2,743 km^3 in 2008 to 3,858 km^3 in 2050 [FAO 11a, FAO 11b]. Much of the increase in water needed for irrigation will be in regions already suffering from water scarcity. To study the impact of water scarcity, there are several simulation models available, for instance GCM (general circulation model) or RCM (regional climate model). In the case of Spain, some of these models, such as AR4 (Rossby Centre regional atmospheric model) and HIRLAM (High Resolution Limited Area Model), predict an increase of 1°C in temperature and a decrease of 5% in precipitation in 2020, and as a result, the decline in water resources from these areas is 10% according to AEMET (Agencia Estatal de Meteorología) [AEM 09]. These values could increase significantly according to these studies, which calculate a decrease of 20% in precipitation and increase of 3°C in temperature by 2050 according to the PNACC [PNA 11].

In Figure 1.1, we show the consequences of water scarcity in two scenarios for several areas in Spain. The first scenario represents simulations with an increase of 1°C in temperature without change in precipitation, and in the other scenario with a decrease of 5% in precipitation.

Taking these estimations into account, the main challenge facing the agricultural sector is making 70% more food available with less water resources. More water resources or more efficient use of resources will be needed to fulfill this objective. We cannot waste water resources and therefore we are forced to improve current irrigation techniques.

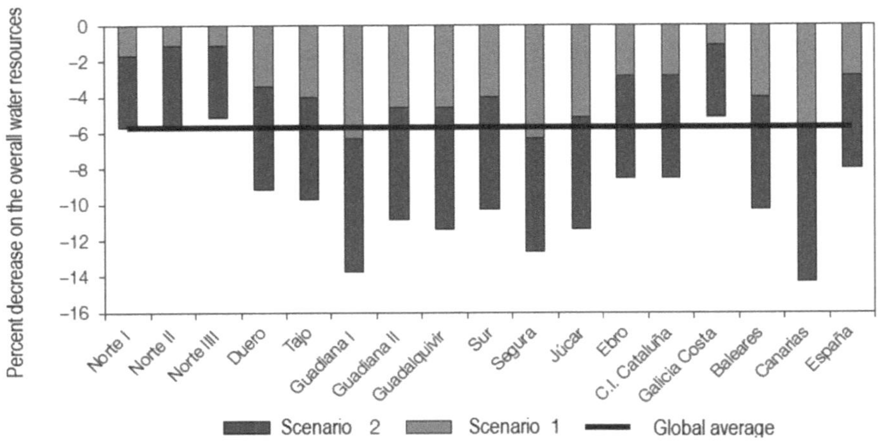

Figure 1.1. *Evolution of the percent decrease for overall water resources in Spain [MIN 00]. For a color version of the figure, please see www.iste.co.uk/bonet/canals.zip*

1.1. Definition of the irrigation problem

The operational aspects of farm irrigation and water supply systems, in areas still dominated largely by tradition, do not usually reflect a high degree of water use efficiency as a primary objective. This efficiency, expressed as the ratio between the quantities of irrigation water effectively used by the crops and the total quantities supplied, has only been considered as an important factor in irrigation over the last 60–65 years. This is not surprising because, up to 80 years ago, our knowledge of the water requirements of crops, more specifically with regards to evapotranspiration, was only vague and investigations on the water resources of irrigated areas were not receiving as much attention as today. Moreover, water was available and its price was low or even free.

There are three physical characteristics that govern any irrigation operation, in terms of both quantity and time:

– the evapotranspiration of the various crops cultivated, which changes during the growing season;

– the moisture content of the soils between field capacity and a preselected depletion limit (that is, the lowest acceptable moisture content that does not significantly affect yields);

– the infiltration rate of the relevant soils.

Other physical factors such as rainfall distribution, topography and canal seepage may, of course, play a role, but the above three characteristics must be considered under all circumstances.

Together, all these factors must serve as the basis for defining such operational features as depth, duration, and interval of irrigation for different crops and soils. But even with this information available, it is only possible to predict the overall irrigation efficiency within an accuracy of 15% [BOS 90, BOS 05].

The lack of basic knowledge of water use efficiencies has several serious drawbacks:

– due to the low efficiency in the planning and design of irrigation systems, a large safety margin is applied;

– investments are thus considerably higher than would otherwise be necessary;

– the limited water resources are not optimally distributed and used, a great deal of water is wasted and fewer areas can be irrigated;

– the low overall irrigation efficiency creates harmful side effects such as rising groundwater tables and soil salinization. To control the groundwater table, a costly subsurface drainage system may be necessary and this will seriously affect the cost of the project.

These irrigation problems are related to the efficiency of water uses and old infrastructure which is responsible for water losses due to cracks in the concrete of the canal side walls.

1.1.1. *Conditions involved in reducing the efficiency*

According to Oca [OCA 09], there are three ways to improve the efficiency of water used for crops: the first is to improve the efficiency of the application of water to the crop by the farmer, the infrastructure and the modes of *water distribution* used by the irrigation communities to transport and distribute the water; the second is to act on *water pricing*, after analysis of the socio-economic effects of increasing or decreasing the water price, in order to encourage the farmers to save water; a third area that we cannot ignore is the *structure of agricultural production* in irrigated areas, because if the profit margin is low, the farmer does not have the chance to adopt new technologies which improve the efficiency of water use.

These three areas are not independent. Although we can build infrastructures to improve the efficiency of water distribution, the farmers must adopt new technologies to make good use of water resources and save money, and all of these things are strongly conditioned by the productive structure.

1.1.1.1. *Water distribution: rotational irrigation scheduling or on-demand irrigation scheduling*

The water distribution from turnout structures requires infrastructure to transport the water; however, the final water distribution to the farmer depends on complex organizational structures that can adopt different forms of distribution. The water distribution system can influence the technique of irrigation and can be used to encourage water saving.

The irrigation communities are usually responsible for water distribution and they also manage the maintenance of canal networks: the main and secondary canals. The irrigation communities are institutions with a long tradition in Spain and in other countries, and their role is the distribution of irrigation water. The oldest democratic governance institution in Europe designed to guarantee the correct functioning of a complex network of irrigation canals and safeguard the interests of irrigation communities is the Valencia Water Court (*Tribunal de las Aguas de Valencia*). This court settles conflicts between irrigation users on an impartial basis and is made up of democratically elected farmers, experts in uses and customs and fair in their proceedings.

The Valencia Water Court (Figure 1.2) has survived over the centuries to the present day and was integrated into the Spanish legal system with the same guarantees and legal value as any civil court. This is due to its effective contribution to the maintenance of the vast and complex system of irrigation canals used for Valencia's fertile plains, built in the age of Muslim Spain (9th–13th Centuries). This highlights the great importance of the water distribution system in Spain due to the dry climate of the country.

Figure 1.2. *The Valencia Water Court. This picture was painted by Bernardo Ferrándiz in 1865 and it is preserved in the Salón Dorado Grande del Palacio de la Generalidad Valenciana (Spain)*

In a general way, we can classify the modes of water distribution depending on the degree of flexibility in water use by farmers.

Rotational irrigation scheduling is the most restrictive of all irrigation systems. The rate, frequency and duration are fixed by the policy of the central water authority and remain fixed for the entire irrigation season. There are a number of possible variations of rotational irrigation scheduling that allow farmers to better fulfill their crop needs. Several methods are presented as follows:

– Continuous flow schedules are a special case of rotation systems, where the duration is a fixed period, the rate is constant and the frequency remains fixed during the growing season. Rotation and continuous flow systems are used in places with plenty of water where the growing season is short and the economics of efficient irrigation for maximum yield is not taken into account. In many cases, while a constant rate is delivered to a farm, the water is rotated between fields or spilled when not needed for irrigation.

– Varied-amount rotation schedules are one way of adjusting the volume of water delivered over different parts of the growing season. In general, the frequency remains fixed, while the duration and/or rate is varied to apply more or less water to a particular area.

– Varied-frequency rotation schedules are another way of adjusting for variations in crop water use over the irrigation season. Under these schedules, the frequency of delivery is varied.

Rotational irrigation scheduling is the most common mode of water irrigation distribution. It is usually linked to an old irrigation infrastructure with a low frequency of water deliveries and a low efficiency for irrigation crops, where irrigation changes are supervised by the irrigation community and not by the user.

The pricing is usually based on the irrigated area of each user, which does not encourage the efficient use of water.

The *on-demand delivery schedule* is the most flexible distribution system. The farmer or the watermaster decides what time and what volume of water should be applied according to the crop needs; several parameters are analyzed and measured for each crop or from the knowledge of the farmer. In effect, in on-demand-oriented systems, we set the irrigation schedule by determining the timing and amount of water irrigation most suitable according to a series of technical procedures. The quantification of the water balance is based on the retention of water in the soil and we can estimate the crop water consumption by monitoring the water status of the soil or plant.

Regarding the monitoring of crop water demand, it should be noted that there is a new generation of sensors used to detect the water needs of the plant with much more accuracy. There are new sensors which use capacitance measurements that allow for the continuous recording of soil moisture at different depths. In this regard, these sensors allow us to detect which are the best periods to irrigate the plant. This has also led to advances in the early detection of water stress of the plant by remote sensing [FER 00].

These different modes of irrigation scheduling have a significant impact on the consumption of water for irrigation and define the degree of development of an irrigation area. In general, the goal is to supply the entire field uniformly with water, so that each plant has the amount of water it needs, neither too much nor too little. There are various types of irrigation techniques which differ in how the water obtained from the source is distributed within the field and these are related to the modes of water distribution.

– In surface irrigation systems, water moves across the surface in order to moisten and infiltrate the soil. Surface irrigation can be subdivided into furrow irrigation (small parallel channels along the field length where the water is applied through each furrow) and basin irrigation (small areas having level surfaces that are surrounded by earth banks, see Figure 1.3). Surface irrigation is often called flood irrigation, as the irrigation results in the flooding of the cultivated land. Historically, this has been the most common irrigation technique in water distribution and this type of irrigation technique is usually used in rotational irrigation scheduling. Among the different modes of rotational irrigation, surface irrigation is the least efficient especially when it comes to shallow soils with low water storage capacity, which usually requires the application of large volumes of water.

– Localized irrigation is a system where water is distributed under low pressure through a piped network, in a pre-determined pattern, and applied in small quantities to each plant. Drip irrigation (Figure 1.5), spray or micro-sprinkler irrigation and bubbler irrigation belong to this category of irrigation system. This type of irrigation technique is usually used in on-demand irrigation scheduling.

Figure 1.3. *Basin irrigation of wheat*

Figure 1.4. *Center pivot with drop sprinklers*

Figure 1.5. *Drip irrigation*

The most important advantage of sprinkler and drip systems is the ability to very accurately modify and adjust the rate and frequency of irrigation; these systems can be easily adapted to the characteristics of soil moisture retention and specific crop needs.

The efficiency of water distribution is not only dependent on the modes of distribution; one of the most relevant causes in loss of efficiency is due to old infrastructure. Waterproof coating may be damaged and therefore the system's conditions can change drastically over time; problems related to water leaks can be significant in many cases or even become excessive.

The on-demand irrigation system needs a larger and more expensive infrastructure than a rotational irrigation system, because an on-demand system has to deliver water as long as the farmer requires it to irrigate the crop. In case of crops with similar water needs or, in particular, periods of water scarcity, the water is delivered to the crops at the same time, increasing the peak flow in the canal. However, on the contrary, the volume of water delivered through on-demand irrigation systems is lower than in rotational irrigation systems, because the application efficiency[1] is at its maximum in on-demand irrigation systems and the farm irrigation efficiency is improved.

1 The application efficiency is the ratio of water that is stored in the root zone for later use by plants to total water applied.

1.1.1.2. *Increase in the water price*

The water price is an important tool to improve the efficiency in use of irrigation water, as we can use this tool to provide economic advantages to the users who make efficient use of available water. It is necessary to adjust the water rate to the real water price to raise awareness to the users of the water value. In the case of irrigation water, it is extremely important to analyze some important peculiarities such as the geographic and climatic conditions of the region and analyze the socio-economic effects of increasing or decreasing water prices. We can use this tool in different ways:

– application of a general increase in water prices as a way of saving water;

– establishment of economic incentives to reduce the use of irrigation water, especially in periods of scarcity.

Increases in water prices must be introduced jointly with a particular water model distribution, that is, the irrigation distribution system should be an on-demand irrigation system, as the farmer could be more accurate in water consumption, increasing the application efficiency and potentially saving water and money.

The majority of pricing services set by the irrigation community is based, as we have already shown in the previous sections, on fixing a price per hectare irrigated, regardless of the volume of water supplied. In Spain, the main obstacle to implementing new fee structures is the difficulty of installing flow meters in each parcel. In this regard, developing methods to measure the flow rate delivered would be useful.

However, before the possibility of increasing water prices for agricultural use, we must bear in mind that for most farms, irrigation water represents a large percentage of their average costs of cultivation on top of those of the seeds, fertilizers, salaries of farmers etc. These costs are often subsidized by the government. This problem linked to the structural characteristics of the farms means that the social and economic impact of such measures can be extremely important for farming agriculture.

1.1.1.3. *Productive structure*

In order to review the future prospects of irrigation, it is fundamental to know the structure of productive irrigated farms and, more specifically, the economic dimensions that they have. One way to characterize them is through the gross margin of farms, which is defined as the balance between the monetary value of gross output and the value of direct costs inherent in production. Figure 1.6 shows a graph of the productive structure of farms in Catalonia, Spain, obtained by the National Institute of Statistics of Spain (INE). According to the criteria of INE, among these direct costs are included seeds, fertilizers, pesticides, water, etc., but it does not include labor, nor the maintenance of machinery and facilities.

Note that 82% of farms have a gross income of less than 48,000 euros per year and these farms occupy 46.5% of the irrigated area. It is also remarkable that over 50% of farms, which occupy about 15% of the irrigated area, can only reach 10,000 euros of gross margin.

This production structure presents a serious problem in the adoption of new technologies which would improve the efficiency of water use by the agricultural sector. The small economic size of most irrigated farms prevents the use of new technology because the capital investment that the modernization process requires is high.

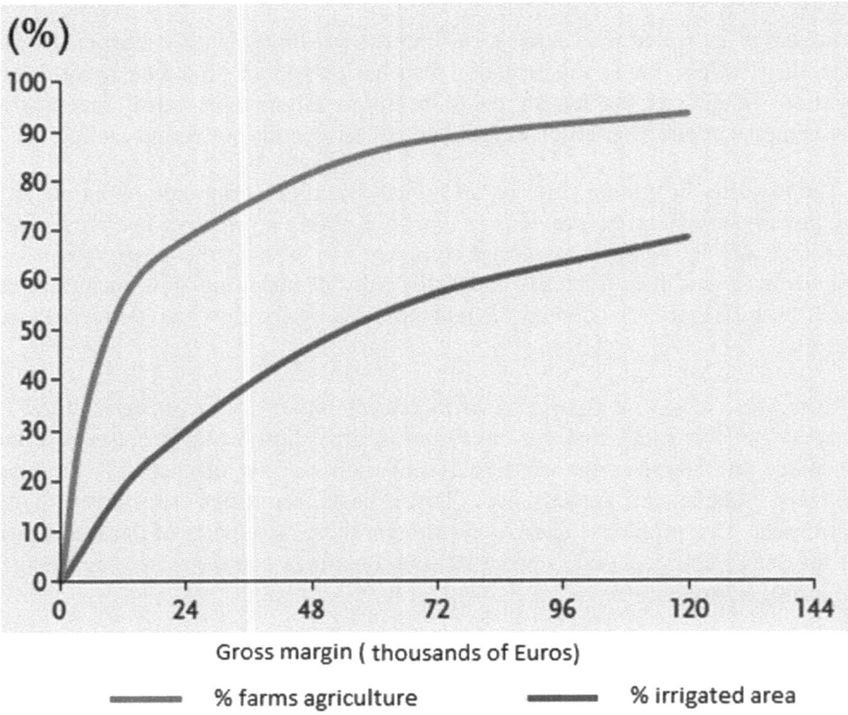

Figure 1.6. *Survey about productive structure of farms during 2007 in Catalonia (INE). For a color version of the figure, please see www.iste.co.uk/bonet/canals.zip*

1.1.2. *Associated structural problems in canals*

The structural problems in canals are frequently associated with bad management which is also related to poor efficiency in water use. The problems in a canal are sometimes a consequence of unscheduled water deliveries or unknown extracted flows by the farmers. Small oscillations in the water level at a cross-section are not relevant, but if the water level variations are quick and higher than a meter, these oscillations could cause serious structural problems in a canal, depending on the water level fall at the cross-section, the height and angle of the side walls and the hydrologic features of the adjacent land.

For instance, in cases where the soil in contact with the cross-section is a non-porous soil, the soil cannot quickly release the pressure in its pores. When there are significant changes in the water level at a cross-section, the gradient of pressures behind the side walls can be so high than the water pressure is able to crack the concrete (Figure 1.7) or even the structure could fail. Put simply, fissures increase water losses and they also reduce the system's efficiency.

Frequently, there are unscheduled water deliveries in a canal and it is not usual to keep records of increments and decrements in the water level. It is therefore quite difficult to know the water level variation in each cross-section along the canal. In this case, it would be useful to have a control element that is able to quantify the flow rate changes in a canal, from water-level measurements at particular cross-sections, to prevent significant water-level oscillations in the canal.

Figure 1.7. *Breakage of the cross-section of a canal from changes in the gradient of pressures (Chimbote canal (Peru))*

1.2. Evaluating irrigation efficiency

There are three areas in which we can act to increase irrigation efficiency:

– changing the water distribution from rotational irrigation scheduling to on-demand irrigation scheduling;

– adapting water prices to the real cost and establishing economic incentives;

– improving the productive structure.

The first and the third point are targeted at introducing new technology, such as canal controllers, and investing in irrigation infrastructure. If we look at the first point, the use of local manual control rather than canal controllers, to manage the demand deliveries in a canal, causes more disadvantages than advantages. It is generally quite difficult to control a canal with local manual control. Simultaneous gate operations cannot be accomplished by local manual control unless irrigation system operators are stationed at every check structure so they can communicate with each other when flow changes occur. On the contrary, we cannot expect the operating staff to control the interaction between all the parameters of a complex system.

Another shortcoming of manual control is the availability of operating staff, as they cannot be continually in attendance at every structure, their degree of dedication and motivation is not unlimited nor is their ability to resist possible pressures from the farmers. As a result, control structures are frequently set incorrectly, and operation is irregular and/or inappropriate. Similarly, measurements and readings may be imprecise. These shortcomings and the resulting lack of reliability, flexibility and equity in the deliveries are responsible for reducing the water distribution efficiency and appear to be the main cause of the very low performance of most irrigation projects.

In spite of these difficulties, we can achieve a satisfactory result between water requirement and demand deliveries, as has been demonstrated using appropriate engineering techniques, and more recently electronics and computer science (ICT), to reinforce traditional practice (Figure 1.8). The use of electronic means is becoming increasingly common in all areas because it is cheaper. For this reason, nowadays it is easier to adapt them to the irrigation crops.

The key elements of any effective control system will include:

– new techniques to estimate the crop water consumption;

– the method of water distribution, such as on-demand irrigation scheduling;

– equipment such as sluice gates, gauges to measure water levels and canal controllers allowing us to control and modify the flow rate or the flow through the delivery turnouts.

The degree of satisfaction of farmers will depend on the productive structure which can be optimized by irrigation techniques and the water distribution system.

Figure 1.8. *The ICT office of the master station of Navarra canal (Spain)*

1.3. Diagnosis and prognosis of the irrigation problem

The stage of diagnosis and prognosis consists of the analysis of the irrigation system to identify shortcomings (diagnosis) and to propose corrective actions (prognosis).

Many algorithms, in particular open loop algorithms, have difficulties and failures when implemented on real canals [ROG 98]. These difficulties are due to deviations between the predictive/control model and reality, and these deviations are even more important when the factors introduced in the following are more relevant:

– conceptual errors in the internal model: the model is based on the Saint-Venant equations, and these equations are only applicable in particular cases, when the pressure distribution along a vertical axis is the same as in hydrostatic conditions, the vertical accelerations are negligible, the curvature of the free-surface is small, etc.;

– errors in the values of physical parameters of the canal: Manning roughness coefficient, discharge coefficients in weirs and gates, etc.;

– disturbances introduced into the canal: infiltration losses by cracks in the coating, unknown withdrawal flow rate, etc.

The conceptual errors associated with the model (the first factor) are not important in canals because the distribution of pressure in a cross-section may be assumed to be hydrostatic, the vertical accelerations are negligible and the curvature of the free-surface is usually small.

Instead, the errors in the values of physical parameters (the second factor) are common in canals, because it is quite difficult to obtain an accurate value of the Manning coefficient or the discharge coefficient in weirs or gates. The model is very sensitive to these physical parameters; for this reason, we have to calibrate them accurately.

The third factor (disturbances introduced into the canal) leads to important deviations between the model and reality. The disturbances caused by infiltration can be controlled by a good coating and these disturbances are not usually important as they have a constant value. Due to the difficulty of distinguishing between disturbances and small perturbations caused by conceptual errors of the model or water infiltration losses, all of them are considered to be the same type of disturbance, which is not important. However, the disturbances introduced by changes in demand deliveries are more dangerous, because the canal could overflow or could dry out depending on the disturbance. These disturbances are usually unscheduled demand deliveries which are caused by flow rate extractions during the irrigation cycle.

As previously mentioned, if the farmers can adjust the supplies obtained by the irrigation community before the irrigation cycle begins, these changes in demand deliveries are known disturbances. These disturbances are known in advance, and these can be mitigated by an open-loop controller (feedforward controller) such as GoRoSo [SOL 03].

The main problem in a canal is the disturbances caused by climatic variations (rainfalls and associated runoff) or unscheduled demand by farmers (which extracts more or less flow than has been demanded by them) because these disturbances are more difficult to mitigate by a controller. One way to protect the canal from these disturbances could be with ponds built by the irrigation community or farmers themselves. A reservoir is able to store water according to the crop requirements and regulates the volume of water provided by the canal, but it is not always possible to build large reservoirs to regulate each pool of the canal. Another option to control the disturbances in canals, which operate in steady state, would be to use the wedge storage of every pool, but this only works with very low disturbances. In all other cases, you need a closed-loop controller (feedback controller) to modify the sluice gate trajectory or a control variable of the system to return to the desired canal state in real time, such as the control algorithm proposed in this book.

1.4. Objectives

In this chapter, we have identified the majority of problems concerning irrigation crops and have proposed some actions to avoid these problems. In this regard, we will focus on a certain action (the introduction of control algorithms) because we have contributed to the development of control algorithms in this book.

Once we have introduced the irrigation problems in transport and water distribution in a canal, we can introduce the necessity of adopting control algorithms in canals to increase the efficiency of water supply and solve some of these problems. This is one of the objectives of this book.

The main objective of this book will be to develop a feedback control algorithm to control the unknown demand deliveries in real time.

2

State of the Art of Canal Control Algorithms

2.1. Introduction

P.O. Malaterre presented a working collection of all of the documentation on canal control algorithms existing at the time, published in the *Journal of Irrigation and Drainage of the American Society of Civil Engineers (ASCE)*; see Malaterre *et al.* [MAL 98]. The great work developed by Malaterre, defining a complete classification of the algorithms involved in canal control, has helped me to structure this chapter. The documentation of the literature review focuses on defining several classifications of control algorithms from different points of view and presents most of the control algorithms that have been developed so far.

We define a control system as an arrangement of algorithms and electronic, electrical and mechanical components which commands or directs the regulation of a canal system. The control system is responsible for controlling structures in a canal based on information extracted from canal measurements, as well as certain variable operating conditions which are defined as targets, and are usually constant water levels. The term "canal control" describes those steps necessary to ensure the required pool water level and flow along the canal. The flow conditions are controlled by adjusting the gates' positions.

Canal conditions depend on the adjustment of action variables (gate openings, discharges), which provide some control over the canal system. Some variables are easily adjusted while others are more complicated or cannot be adjusted. One variable which is easily adjusted is the gate position. Other variables, such as the Manning roughness and the geometry of the canal, cannot be readily controlled. Canal control is achieved by manipulation of the variables to obtain the desired canal system conditions.

The control algorithm, which contains the set of rules or set of instructions, is one of the primary elements in a controller, and it calculates the desired value (position or speed) that the gate must fulfill. An algorithm is a prescribed set of well-defined rules or processes for the solution of a problem in a finite number of steps. In canal control, algorithms are the procedures used to implement the various canal methods of operation. The algorithm is designed to process the input information from the sensors, perform the comparator function and calculate the proper output to the actuator. The input is the quantities that are observed, measured or predicted; the output is a control action.

There are many criteria with which to characterize canal control algorithms, such as control variables (flow, depth, gate opening), interactions between the local controllers (local, centralized, etc.), communication requirements, technical design, alarms and localization of measurements along the canal.

Different control algorithms for the regulation of irrigation canals have been developed and applied throughout the world. Each of them can be characterized according to several criteria, among which are: the considered variables (controlled, measured and control action variables), the logic of control (type and direction), the control system design and the field implementation.

2.2. Considered variables

In control theory, a system is usually represented as in Figure 2.1. “U” represents the control action variables (e.g. gate openings, increments of gate openings, discharges) to modify the state of the system, and “Y” represents the system controlled variables (e.g. water level, discharge, volume) or measured variables (e.g. water level, discharge). There are three kinds of variables in a control algorithm:

– controlled variables

– measured variables

– control action variables.

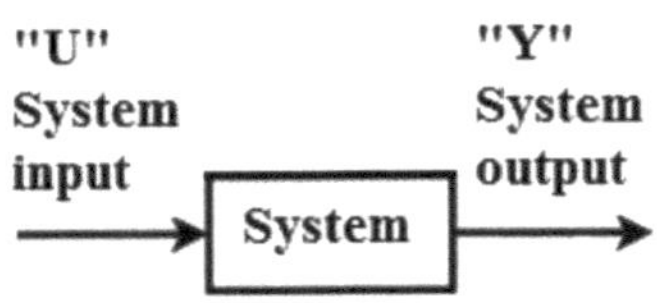

Figure 2.1. *Representation of a system*

2.2.1. *Controlled variables*

Controlled variables are system variables to which targets are assigned. The controlled variables are those describing the desired behavior of the canal. Controlled variables on an irrigation canal can be one of the three types: flows, water levels or volumes, according to the terminology used in the physical system modeling. The controlled variables are localized at certain cross-sections in the canal.

2.2.1.1. *Controlled variable: discharge*

The needs of irrigation canal users are defined mainly in terms of discharge. For example, agricultural needs are expressed in terms of given discharges delivered to a parcel, to a secondary canal or to a pumping station; environmental needs are expressed as tail-end discharge, or minimal discharge; users' needs can then be defined in a more flexible way, in terms of volume distributed over a time period. In the latter case, the controlled variable is no longer a given value of discharge, but a volume.

Discharge fluctuations are then authorized, but these disturbances can be managed by controllers or by a storage reservoir, which is not the most common option due to the fact that reservoirs are expensive and the space available next to the canal could be limited, as we explained before.

A hydraulic system has to be managed, directly or indirectly, in order to satisfy users' demands in discharge. Considering the nature of the physical phenomenon (gravity surface flow from upstream to downstream), these demands in discharge can initially be satisfied by the volume of water in the canal, but these demands are usually satisfied by a source situated at the upstream end of the system, which is often a reservoir. Rodellar [ROD 93], CACG [PIQ 82], CARAMBA [DEL 86], Davis University [BAL 85] and SIMBAK [CHE 91] are examples of regulation methods which control discharges.

2.2.1.2. *Controlled variable: water level*

The water level can easily be measured in canals, so it is usually the controlled variable. Controlled water levels "y" can be upstream (Figure 2.2(a)), downstream (Figure 2.2(b)), between the two or inside the pool (Figure 2.2(c)), and operational characteristics are very different depending on the location of "y". The methods of operation should not be confused with canal operation and control concepts. These methods of operation are based on the location of a pivot point within the canal pool, where the water level remains constant, while the water surface slope varies. The methods of operation are as follows [BUY 91]:

– Constant upstream depth: the pivot point is located at the upstream end of the canal pool (Figure 2.2(a)).

– Constant downstream depth: the pivot point is located at the downstream end of the canal pool (Figure 2.2(b)).

– Constant volume: the pivot point is located near the midpoint of the canal pool (Figure 2.2(c)).

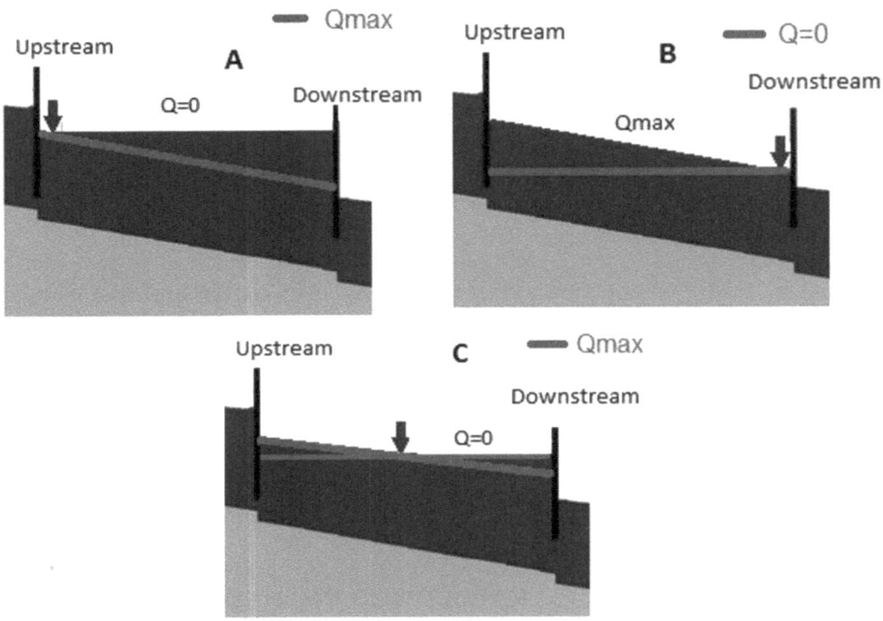

Figure 2.2. *Downstream control concept. For a color version of the figure, please see www.iste.co.uk/bonet/canals.zip*

Before defining the methods of operation, we must introduce several concepts, such as canal operation and canal control, in order to thus make the methods of operation more understandable.

The canal operation concept, established by the canal system operating criteria, determines the flow schedule. We can define the canal operation concept as follows:

– The downstream operation concept applies to canal systems that are primarily demand-oriented and is usually associated with delivery systems. These systems convey water from a single source, such as a storage reservoir, to a number of individual points of use, and they are associated with irrigation. The delivery systems initially convey the maximum flow and it decreases downstream.

– The upstream operation concept applies to canal systems that are primarily supply-oriented and the concept is usually associated with collector systems. These systems convey water from several individual sources, such as surface inlet drains from a rainstorm located upstream, to a single point of diversion located downstream. The collector system is supply-oriented because it must convey a variable water supply from any number of sources to a downstream single point of diversion. The collector systems initially convey the minimum flow and it increases downstream.

The control concept determines how the canal control structures are adjusted to satisfy the canal operation concept. The control concepts used in a canal system are defined by the location of the information needed in order to operate the control structure:

– *Downstream control*: control structure adjustments are based on information from a pool downstream. The required information could be measured by a sensor located downstream or by the irrigation system operators, and the watermaster establishes the water schedule from the required information. Downstream control transfers the downstream canalside turnout demands to the upstream water supply source, and is compatible with the downstream operation concept.

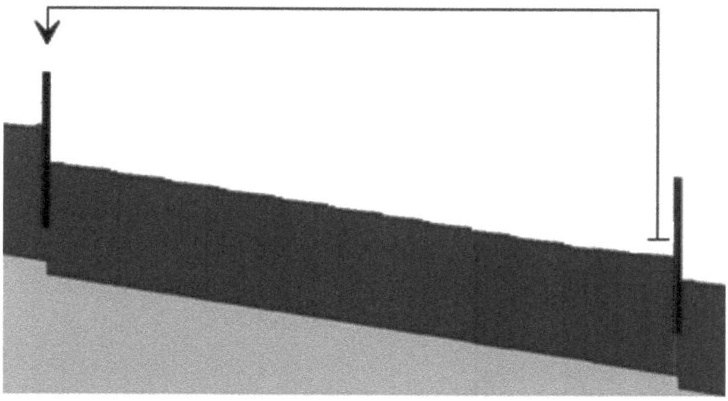

Figure 2.3. *Downstream control concept*

– *Upstream control*: control structure adjustments are based on information from a pool upstream. The required information could be measured by a sensor located upstream of the position of the gate/regulator, and the watermaster establishes the upstream water schedule from the required information. Upstream control transfers the upstream water supply information (or inflow) to downstream regulators, and is compatible with the upstream operation concept.

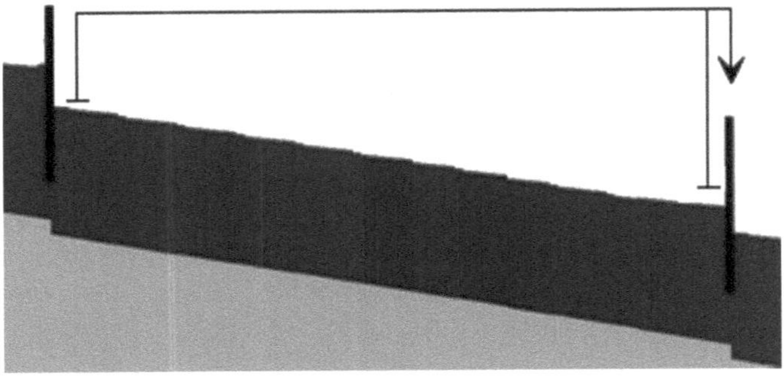

Figure 2.4. *Upstream control concept*

Once we have defined the control concept and the operation concept, we can define the methods of operation.

Constant downstream depth

The constant downstream depth method of operation, wherein the water depth at the downstream end of each canal pool remains relatively constant, is used in most canal systems. The reason why this method is so prevalent is that a canal can be sized to convey the maximum steady flow (Figure 2.5); steady-state water depths should never exceed the normal depth for the design flow rate. The cross-section size and freeboard can be minimized, thus reducing construction costs.

When a constant depth is maintained at the downstream end of the canal pool, the water surface profile will essentially pivot about this point in the canal flow, as shown in Figure 2.5. A storage wedge between different steady-state flow profiles is created. When the flow increases, the water surface gradient and storage volume must also increase. Conversely, the storage volume must decrease for a reduction in steady-state flow.

Owing to these storage considerations, there is a natural tendency for a flow change that originates at the upstream end of a pool to create the change in storage that is needed in order to keep the downstream pool depth constant (Figure 2.6).

Examples of regulation methods controlling downstream water level are GoRoSo [SOL 03]; PI controller (Litrico X. *et al.* [LIT 06] and Aguilar *et al.* [AGU 12]); LQ control [WEY 03]; Model Predictive control [VAN 06]; KAPTROLLER [AKO 95]; IMTA-Cemagref, tested in Yaqui canal (Mexico) [CHA 94]; PRECOM [MAL 96]; Buyalski and Serfozo [BUY 79]; Deltour [DEL 92] and Chevereau [CHE 91]. There are other regulation methods that combine several controlled variables (discharge and downstream water level), such as CLIS [LIU 98], or use a distributed model predictive controller, such as Alvarez *et al.* [ALV 13] or Rodellar *et al.* [ROD 93].

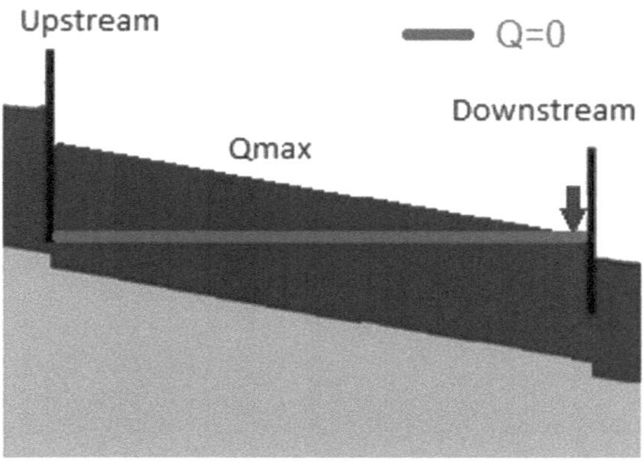

Figure 2.5. *Control of the downstream level of a pool*

Figure 2.6. *Upstream flow change with the constant downstream depth method of operation. For a color version of the figure, please see www.iste.co.uk/bonet/canals.zip*

Constant upstream depth

With this method of operation, a constant upstream depth is maintained by pivoting the water surface at the upstream end of the canal pool, as shown in Figure 2.7. The constant upstream depth method is sometimes called a "level bank" operation, because canal banks must be horizontal in order to accommodate the zero-flow profile (Figure 2.7). The construction of a level bank canal is the main problem of this method. A level bank canal increases the cost of construction considerably, especially for concrete-lined canals, because this cross-section is not constant along the canal. Most existing canals cannot use a level bank operation unless additional bank and lining are added to the downstream portion of each pool. Exceptions to this would be canals with little elevation drop between checks, or those operating at flows well below the maximum flow capacity.

Flow changes originating at the downstream end of the pool cause canal water depths to change in the direction that is needed in order to achieve new steady-state profiles. AVIS and AVIO gates [GOU 93] and PIR [DEL 92] are examples of regulation methods which control upstream water level.

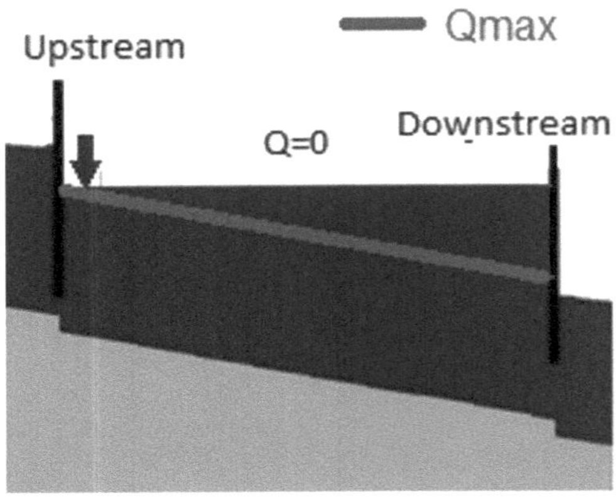

Figure 2.7. *Control of the upstream level of a pool*

Constant intermediate water level

This method of operation is based on maintaining a relatively constant water volume in each canal pool at all times. The water surface will pivot about a point near midpool as the flow changes from one steady state to another.

There are storage wedges on either side of the midpool pivot point, as shown in Figure 2.8. For any given flow changes, volume changes in each of these wedges are equal and opposite. When the flow decreases, the volume of water decreases in the upstream wedge and increases in the downstream wedge. When the flow increases, the opposite occurs.

With the constant upstream depth and constant downstream depth methods, an excessive amount of time is required in order to either build up or deplete the storage in the entire canal system when changing the steady-state rate of flow. The constant intermediate water-level method of operation avoids lengthy delays, because the total volume of water in the canal system does not change significantly. BIVAL [ZIM 87] is an example that demonstrates regulation methods controlling constant volume.

On the other hand, canal dimensions must be adapted to the desired demands, and as a consequence, oversized canals may be finally necessary in comparison to other procedures. This method of operation is rarely used.

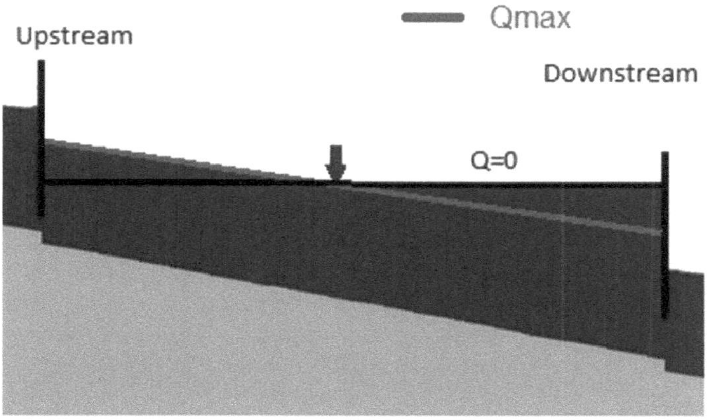

Figure 2.8. *Control of an intermediate water level*

2.2.1.3. *Controlled variable: volume*

The methods which use the volume as a controlled variable offer the advantage of drastically reducing transients by making the flow rate variations in each pool independent of the stored volumes. These methods are applicable to irrigation canals with important storage volumes, and they are equipped with turnouts whose feeding is not dependent on the water levels in the main canal (e.g. pumping stations). Some examples of these methods are Dynamic Regulation [COE 77], Controlled Volumes [BUY 91], BIVAL [ZIM 87] or "A multivariable approach for the command of Canal de Provence Aix Nord water supply subsystem (MIMO)" [VIA 04].

2.2.2. *Measured variables*

Measured variables on irrigation canals are generally water levels (e.g. ELFLO [SHA 71], PIR [DEL 92], PREDAPTROL [ROD 93], PRECOM [MAL 96], KAPTROLLER [AKO 95], Corriga [COR 82] and GoRoSo [SOL 03]). In some cases, measured variables can be discharges (e.g. CACG [PIQ 82] and CARAMBA [DEL 86]). A discharge can be measured with specific equipment (generally based on the measure of one or several flow velocities, with a propeller, an ultrasonic device or an electromagnetic device), through a cross structure equation or a local control section rating curve with sufficient precision. A depth can be measured by installing pressure transducers, bubbler pipes or gages directly in the cross-section.

Automatically controlled canal operation depends on water-level data collection. Accurate water-level information is critical as the actions taken concerning the canal are based on these water-level measurements.

2.2.3. *Control action variables*

Control action variables require mechanical equipment such as gates, valves and pumps.

Control action variables "U" are generally either gate openings or increments of gate openings, and "Q" are generally discharges or increments of discharge. Gate openings have the advantage of taking into account the complex dynamics linking this opening with the local discharge and upstream and downstream water levels. Below are some examples of controllers that use these control variables.

These variables are the gate positions (G) and the discharge (Q):

– the discharge: the control action is the flow rate (Q), and it requires a convert function to translate it into the gate position (G), so that it can be applied to the system. Examples of this case would be CACG [PIQ 82] and KAPTROLLER [AKO 95];

– the gate position: this has the advantage of taking into account the complex gate dynamics. Controllers that use the gate position as a control action variable are CLIS [LIU 98] and Sepúlveda [SEP 07].

2.3. Logic of control

As it depends on the type of information used to calculate the control action variables of the system, to define the logic of control concept, we will talk about "closed-loop" (also known as "feedback control"), or "open-loop" (also called "feedforward control").

2.3.1. *Open-loop control*

Open-loop control is when the controlled variable is adjusted with no comparisons to actual response or to actual desired conditions. An example of open-loop control is a traffic light signal. The traffic lights respond at specific time intervals regardless of the actual traffic conditions.

In open-loop, the control action "U" is calculated knowing the dynamics of the system (using a model), the setpoint output "Yc" (for instant water demands by farmers, known in advance) and possibly an estimation "P" of perturbations (Figure 2.9). The open-loop can compensate for inherent system time delays by anticipating users' needs. These needs must be known as precisely as possible. They should take into account climatic, agronomic and sociological data, as well as recordings of the water consumption of previous weeks or seasons [PER 89]. An open-loop is generally insufficient due to model errors, perturbation estimation errors and unknown perturbations. Open-loops can be applied to all of the controlled variables: discharge, water level and volume. Some examples of such methods are Bautista *et al.* [BAU 03], Baume [BAU 93], Wahlin & Bautista [WAH 03], Soler [SOL 03], Sawadogo [SAW 91], Sabet *et al.* [SAB 85], Tomicic [TOM 89] and Huising [HUI 04].

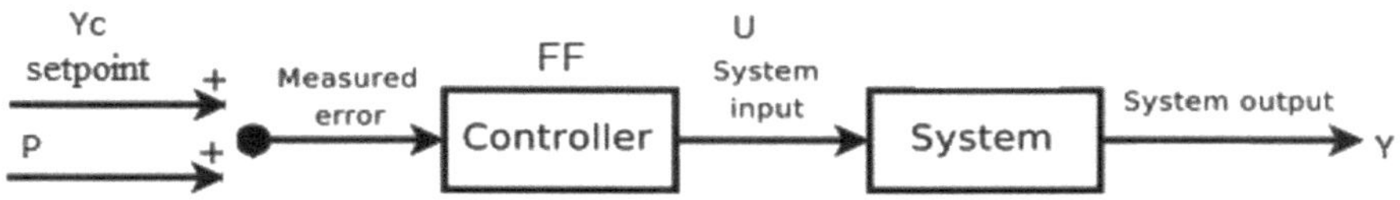

Figure 2.9. *Open-loop control (feedforward)*

2.3.2. *Closed-loop control*

Closed-loop control is when the controlled variable is measured and compared with a setpoint representing the desired performance. Any deviation from the setpoint is sent to the control system, so that it can set the control actions to reduce the deviation of the controlled variable from the reference.

In closed-loop, the control action variable “U” is calculated from the error measured between the real controlled variable “Y” and its corresponding setpoint “Yc” (Figure 2.10). Perturbations “P”, even if unknown, are taken into account indirectly through their effect on the output “Y” of the system. In control theory, this concept is essential since it links a control action “U” to a controlled variable “Y”. Closed-loops can be applied to all of the controlled variables: discharge, water level and volume.

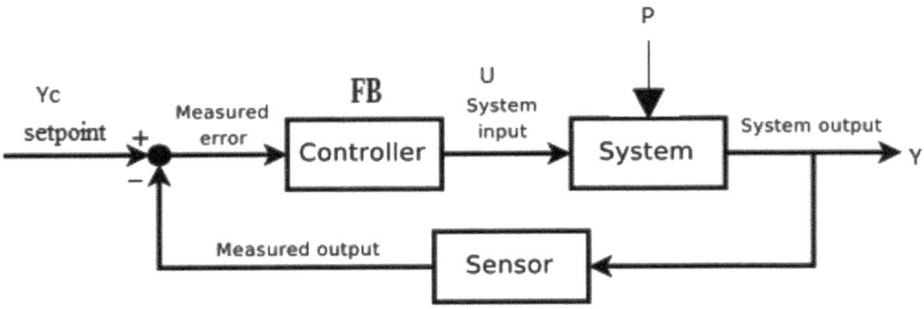

Figure 2.10. *Closed-loop control (feedback)*

Examples of closed-loops in discharge are GPC [SAW 92], Rodellar *et al.* [ROD 93], CACG [PIQ 82], CARA [MAR 89], CLIS [LIU 98] and Isapoor *et al.* [ISA 11].

We can introduce two points of view about the closed-loops in water level depending on the relative locations of the control action and controlled variable, as was defined in the canal operation and the canal control concept.

Downstream water-level control indirectly generates a discharge closed-loop control, since it is obtained from the modification of the upstream discharge. This characteristic is an essential property of the downstream water-level control. Some examples of such methods are AMIL, AVIS and AVIO gates [GOU 87], Little-Man downstream and ELFLO [SHA 71], CARDD [BUR 83], Zimbelman [ZIM 87], PID [CHE 91], PIR [DEL 92], MPC [VAN 06], Clemmens & Schuurmans J. [CLE 04] and Gómez *et al.* [GÓM 02].

Some water-level control methods combine the logic of upstream and downstream controls. These are called mixed feedback controls. Some examples of such methods are Balogun [BAL 85, BAL 88] and Filipovic *et al.* [FIL 89].

2.4. Methods of control system design

The design technique is the algorithm or methodology used within the control algorithm in order to generate the control action variables from the measured variables (i.e. the determination of control action "U" leading to the desired dynamic behavior of the controlled system).

2.4.1. *Controller design: definition of inputs/outputs*

In canals, discharge is normally the parameter that is to be controlled, although it is not a quantity which can readily be determined. Instead, some other quantity must be measured; then, discharge is related to these quantities through algorithms. In canals, the most frequently used parameter is water surface elevation, or the difference in water surface elevation between two points. Unless stated otherwise, water surface elevation is the controlled parameter (input). The location and number of sensors used as the output depend on the objectives of the conveyance system. Only one water-level sensor per reach is used, if the canal is a delivery system consisting of gravity turnouts. Usually, the location of the sensor is near the downstream end of a canal reach near the turnouts. Multiple sensors in a canal pool are used to implement various control methods.

Control action variables "U" and controlled variables "Y" of a given controlled system are defined in section 2.1. Variables "U" are particular inputs of the system and "Y" are particular outputs of the system. Since our classification concerns controllers and not systems, we choose to define "inputs" and "outputs" in reference to the controller and not to the system. The different types of inputs and outputs are illustrated in Figure 2.11, where:

– U' are variables acting on the system and not generated by the controller (e.g. non-controllable inputs, perturbations);

– U are variables generated by the controller and acting on the system (e.g. control action variables);

– Y' are variables generated by the system and not used directly by the controller;

– Y are variables used by the controller and generated by the system (e.g. measured variables used by the controller).

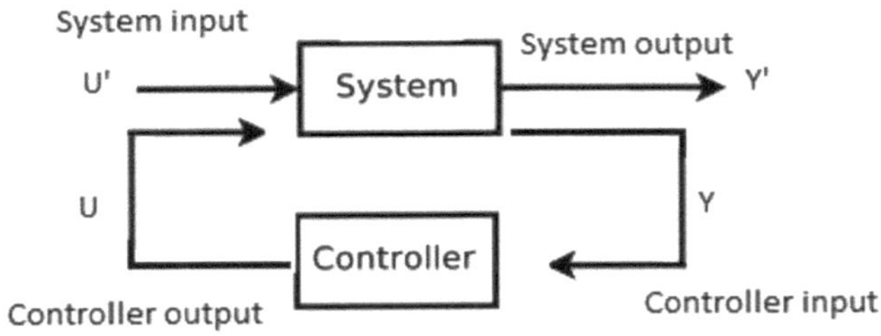

Figure 2.11. *Inputs / outputs of a controlled system*

A monovariable controller (in input and output) has only one input and one output (variables being considered as scalars and not as vectors). Monovariable methods are also called SISO (Single Input, Single Output). A multivariable controller (in input and in output) considers several inputs and several outputs. The multivariable methods are also called MIMO (Multiple Inputs, Multiple Outputs).

2.4.2. *Monovariable methods*

These methods are traditionally used in irrigation canal control. Monovariable methods are applied to a single pool of the canal; they require the overall system to be split into several subsystems without explicitly taking into account any interactions between them. An irrigation canal is a multivariable system presenting strong interactions between subsystems. For example, the operation of a gate influences several upstream and downstream pools.

Most irrigation canal control methods based on control theory use the well-known linear monovariable PID controller. Some examples of PID-related methods are:

– P: AMIL, AVIS, AVIO [GOU 87];

– PI: ELFLO, BIVAL [LIT 03];

– PID: PID UMA Engineering.

Although very efficient in most cases, PID controllers do not explicitly take into account the characteristic time delays in a canal. Some examples of these methods are IMTA [CEM 94], BIVAL, ELFLO and PID UMA Engineering.

There is also the generalized predictive control method (GPC), a monovariable optimization method based on the minimization of a criterion "J", which calculates the control action variable and the error between the controlled variable and its targeted value. This method was developed by Sawadogo [SAW 92], Rodellar [ROD 93] and Camacho and Bordons [CAM 04].

The GPC method uses transfer function models (Chan *et al.* [CHA 90], Soeterboek [SOE 90], Lee *et al.* [LEE 90], Linkens *et al.* [LIN 92]) and it incorporates an open-loop and a closed-loop.

There are other algorithms that use fuzzy control methods, such as CNABRL on the T2 canal, a method based on fuzzy control expert systems used in Marrakech (Morocco) or methods based on neural networks (e.g. Schaalje *et al.* [SCH 93] and Toudeft [TOU 94]).

The Little-Man is another controller that uses a monovariable method, which was implemented in The Columbia River Basin Project (near Ephrata, Washington). The Little-Man control technique is applied to canal systems using the local automatic control method. The application of the Little-Man algorithm is most successful in implementing the constant upstream control method of operation.

Monovariable methods take into account just one variable of the system, without explicitly taking into account any interactions between them. On the contrary, an irrigation canal is a multivariable system that presents strong interactions between them.

2.4.3. *Multivariable methods*

A multivariable controller is designed to control several pools, and therefore, the controller operates globally in the canal, unlike monovariable controllers, which are designed to control a single pool.

Multivariable controllers for canal automation have been designed based on optimal control techniques: linear [RED 96] or nonlinear optimization and model inversion [LIU 95].

Different model inversion methods are described in the literature, leading generally to open-loop controllers such as Chevereau [CHE 91] and Liu *et al.* [LIU 92], and more rarely to closed-loop controllers [LIU 94].

Optimization methods have also been developed. These methods are, in essence, multi-variable. Several different methods exist: linear optimization [SAB 85]; nonlinear optimization, such as Tomicic [TOM 89], Khaladi [KHA 92], Lin [LIN 92], Soler [SOL 03], van Overloop, P.J. [VAN 06], Álvarez A. *et al.* [ALV 13]; LQR [COR 83], Garcia [GAR 92], Filipovic [FIL 89], Reddy [RED 92], Malaterre [MAL 94], Sawadogo *et al.* [SAW 95]; LQG [LEM 13], Wahlin [WAH 04], Montazar *et al.* [MON 05] and PI control [LIT 06], Begovich *et al.* [BEG 05]. The classic nonlinear optimization only leads to an open-loop, sensitive to errors and perturbations. In order to introduce a closed-loop, the optimization has to be processed periodically. This complicates the method and limits its applications due to real-time constraints and computer time. Furthermore, the determination of real initial conditions, required for optimization, is not easy. On the other hand, based on a state space representation, LQR methods incorporate, in essence, an open-loop and a closed-loop.

The implementation of multivariable methods is far more complex than that of monovariable methods.

2.5. Field implementation

In the history of canal control, several multivariable methods have been developed. However, very few of them have been finally implemented on canals.

Different aspects of field implementation of regulation methods can be distinguished. These are the configuration (e.g. localized, semi-localized, centralized), devices (e.g. automatic gates), instrumentation (e.g. water-level sensor, discharge measurement device), communication (e.g. radio transmission), calculation and data processing.

In local configuration (Figure 2.12), each device is controlled by an independent controller. The device (e.g. PID UMA Engineering, Little-Man, GEC Alsthom Gates) only uses local information (measured variables). A local automatic controller consists of all of the equipment required to execute the control algorithm and the necessary inputs and outputs in a stand-alone unit, usually located at each check structure. Some examples of this method of implementation are Gómez *et al.* [GÓM 98], Cardona *et al.* [CAR 97], Begovich *et al.* [BEG 07] and Van Overloop *et al.* [VAN 05].

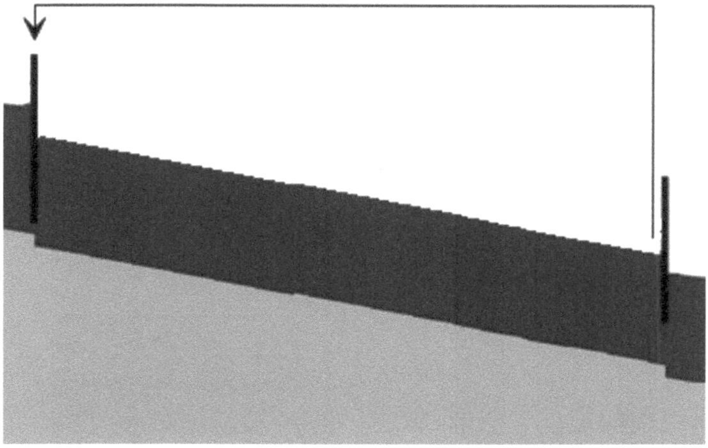

Figure 2.12. *Local controller*

In a semi-local configuration (Figure 2.13), controllers are no longer independent. They explicitly use data generated by adjacent controllers or setpoints. For example, a controller can use the control action variable of the next downstream controller (e.g. PIR, ELFLO + decoupler, semi-decentralized predictive control [SEP 07]).

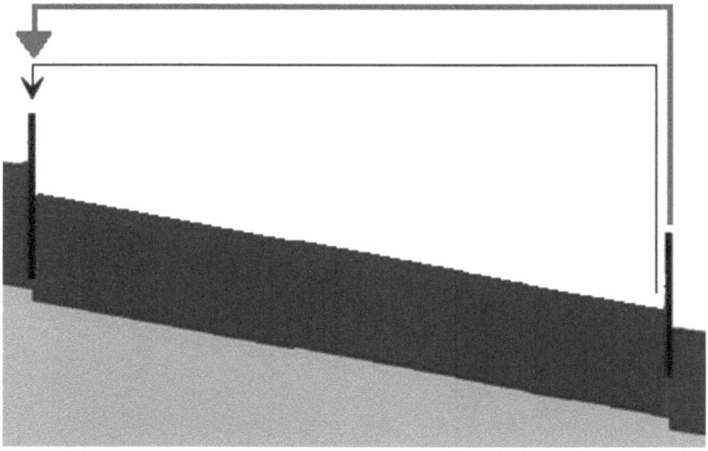

Figure 2.13. *Semi-local controller*

In centralized configuration (Figure 2.14), all control actions are generated by a central controller (human being or computer) in headquarters where information from the whole canal is available and used in the controller, distant from the different actuators. Control actions are gate openings (e.g. the Salt River Project) or discharges. Centralized control allows for supervision and remote control of the system. However, the implementation is more complex and more sensitive to hardware and communication breakdowns than localized configurations. Some examples of this implementation are GoRoSo [SOL 03] and Montazar *et al.* [MON 05].

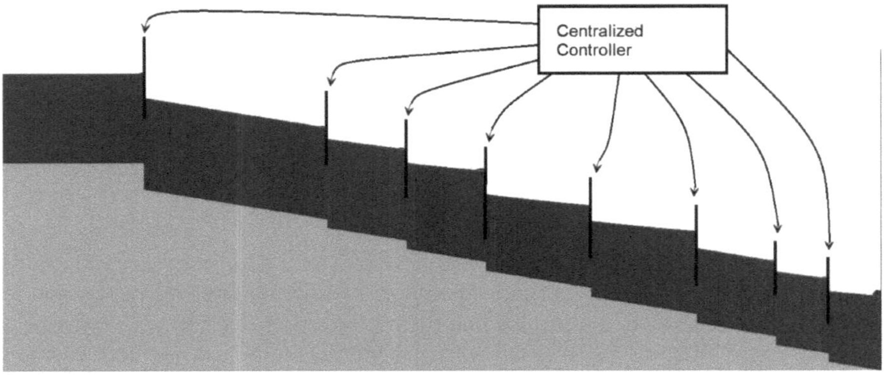

Figure 2.14. *Centralized configuration*

3

Characterization of a Predictive Control Scheme

The term "canal control" describes those steps needed to ensure the required water level and flow along a canal. The control of the canal is achieved through the manipulation of variables (sluice gate openings, pump flows) in order to obtain the desired conditions in the canal. These conditions are controlled by adjusting the volume of water pumped in/out of the canal, adjusting the positions of gates, and by regulating flows through the delivery turnouts. When a canal is in steady flow, the control of the canal is easy, but problems appear due to flow changes at setpoints. The demand deliveries are not constant in space nor in time. The canal state may vary unpredictably due to disturbances, which are considered to belong to the external environment of the system (runoff produced by rain events, canal losses due to infiltration, unknown turnout operations). Therefore, if we want to keep the canal under control in these circumstances, it is necessary to use controllers such as, for instance, proportional integrated derivative controllers [MON 05], predictive controllers [GÓM 02], optimal controllers [MAL 98] or heuristic controllers [DUR 04].

If the canal system is operated by automation, the efficiency, flexibility and responsiveness of the diversion project will be improved, and the water resource loss will also be reduced during the diversion project operation process [BUY 91].

The global state of a canal is often governed by predictive controllers, which modify the gate position depending on the demand or on-demand delivery changes, and these controllers have to take into account a lot of variables that affect the canal flow behavior; parameters such as the Manning or energy loss coefficient, for instance, or the past and current canal state. Before considering a dynamic system (the canal state) governed by an overall control diagram as we will explore in Chapter 4, we will characterize how our predictive controller should work.

3.1. Introduction

The main focus of predictive control (PC) is to define control actions using input data from a computer model, thus defining the dynamics of a canal system that can be correctly approximated by Saint-Venant's equations or by another hydraulic approach and the hydraulic model of a check structure (such as a sluice gate and a weir) [BUY 91, YEV 75]. On the other hand, physical parameters of the computer model (such as the Manning coefficient) must be calibrated before using the predictive control.

We should follow the next steps in the process of a predictive control algorithm:

– estimate the canal flow response during a predictive horizon;

– determine the sequence of control actions to control the canal during a predictive horizon.

Canal control is strongly associated with taking more actions, depending on the reference defined by the watermaster; normally this response is to increase or decrease the water level at a setpoint or checkpoint, that is, at a certain cross-section in the canal in which we can measure the water level with pressure transducers or staff gages directly in the canal. Typically, these setpoints are associated with a certain constant value (desired water level) during the irrigation cycle.

A control action to minimize the variation between the desired and simulated/measured behavior is usually made by optimization problems, which seek to minimize an objective function. Minimizing the objective function is equivalent to minimizing the absolute deviation between simulated values and desired values. Before introducing the optimization problem of our predictive control, we are going to introduce a general definition of computer models and objective functions that are also involved in predictive controllers.

3.2. Computer model

An irrigation canal is a hydraulic system whose objective is mainly to convey water from its source (reservoir) down to the users (farmers). This system is controlled by cross structures such as sluice gates, weirs and orifice offtakes, which are operated in order to control the water levels or discharges in several cross-sections along the canal.

The physical dynamics of this hydraulic system can be correctly approximated by Saint-Venant's equations, which are nonlinear partial derivative hyperbolic equations, combined with nonlinear algebraic cross structure equations depending on the cross structure (weir equation, sluice gate equation). These must be completed using initial conditions and boundary conditions at cross structures and the whole canal.

In irrigation canals, we have an advantage over other fields that use hydraulic modeling due to the regularity and shape of the cross-sections. The flow in the canal can be simulated by 1D flow models. Although this is true in reality, in certain sections such as cross structures, the flow behavior is generally 2D or 3D. Using 1D flow models, however, is very advantageous because these models decrease the computational time compared with 2D or 3D models; the computational time is a crucial factor to take into account for a predictive model. An unsteady 1D flow model is used to analyze the system behavior in this book and the model results are visualized with our displayer (Figure 3.1).

In the literature, there are several models describing flow in canals. Some of these models are established from the full Saint-Venant equations and are called nonlinear models, while other models are established from approximations of the Saint-Venant equations and are called linear models.

Simplifications and approximations of the Saint-Venant equations can be divided into four categories: empirical, linearized, hydrological and hydraulic [FRE 85]:

– empirical methods are based on a large collection of observed data for a discrete reach of a river or canal and are only applicable for the reach;

– linearized methods simplify the Saint-Venant equations by neglecting nonlinear terms;

– hydrological methods utilize the mass conservation equation and a relationship between storage and discharge;

– hydraulic methods add the momentum conservation equation.

However, linear models are the ones commonly used in predictive control. These models simplify the Saint-Venant equations by linearizing them around a steady state, called a reference state [BRO 85]. The advantage of a linear model is the simplicity. However, this same characteristic is responsible for the disadvantages. These models do not consider the nonlinear terms, which have an important influence on the system dynamic when the flow changes are important over time.

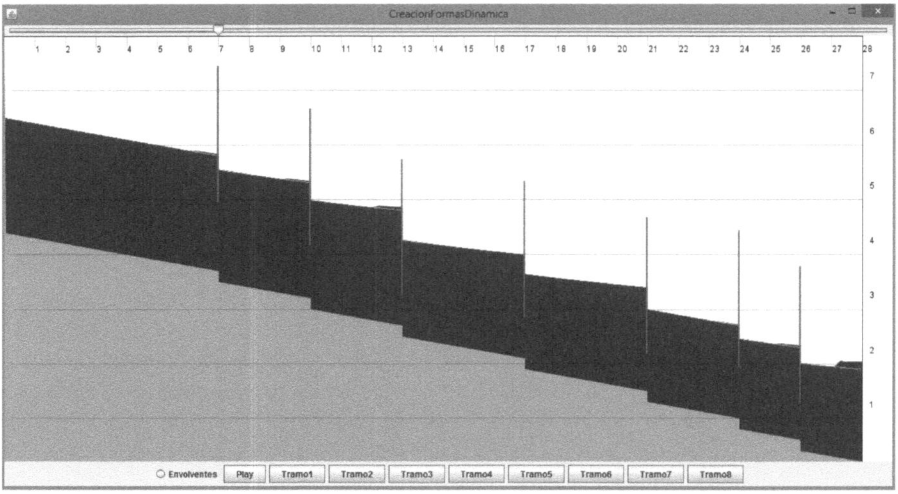

Figure 3.1. *Display of the values obtained by a computer model of a canal (this displayer was programed in Java code and it was developed for us to check the results obtained with predictive control)*

3.2.1. *Linear model*

There are several linear models in the bibliography and we will introduce some of the more common examples:

– The Muskingum model is a frequently used hydrologic linear model for flood routing (Cunge [CUN 69]; Dooge *et al.* [DOO 82]; Viessman *et al.* [VIE 95]). It contains two equations: a continuity and a storage equation. The parameters of the model are K (the travel time of the canal reach) and χ (dimensionless coefficient weighing the relative effects of inflow and outflow on the reach storage), which together contain all the information about the river or canal reach.

– The first- or second-order Hayami model is derived from the diffusive wave equation (Alvarez [ALV 04]; Chentouf [CHE 01]), a simplified form of the full Saint-Venant equations.

– A linearized model was used by Malaterre [MAL 94], based on a linearization of the full Saint-Venant equations (a finite difference approach to the Saint-Venant equations) around a steady state.

We can generalize all these models and give a global point of view of them, as all these models have to represent the state vector, which describes the time evolution of the hydrodynamic variables of a canal (water level and velocity) in terms of control actions. An example of the equations that represent a linear model can be written as follows:

$$x(k+1) = [a_1]x(k) + s(k+1)$$
$$s(k+1) = [a_2]u(k+1) + \cdots + [a_n]q(k+1) \qquad [3.1]$$

The state vector x(k+1) is the state vector at time step (k +1) and it contains a set of n_x elements of water level and velocity at every cross-section. The vector s(k+1) is the input data vector that contains the gate trajectories, extracted flows, demand deliveries, etc., at time step k+1. The matrices $[a_1]\ldots[a_n]$ are constant and contain the linearization of Saint-Venant's equations. The vector u(k+1) contains the gate position trajectories from the time step k to k+1 (see Figure 3.2). The vector q(k+1) contains the changes in flow rate from the time step k to k+1, which is called the extracted flow vector.

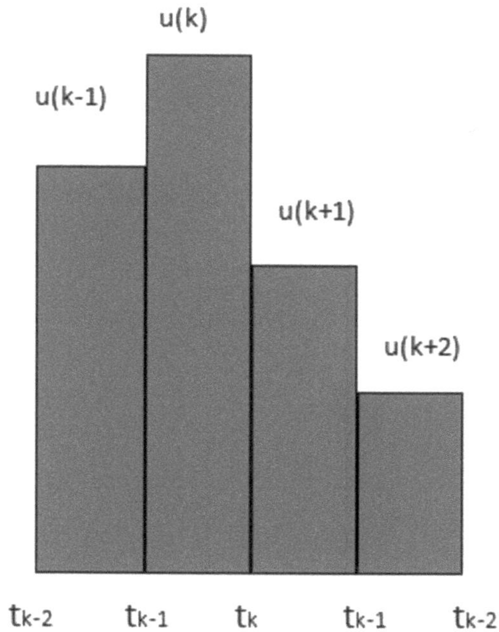

Figure 3.2. *A gate trajectory*

We can define this state vector using a predictive horizon, for instance, between the time steps k_{I+1} and k_F. The vector, which compiles all terms, is called the prediction vector, that is, the output data of the model:

$$X_{k_I+1}^{k_F} = [x(k_I + 1)^T + \cdots + x(k_F)^T]^T \quad [3.2]$$

The same thing happens with the gate position vector ($U_{k_I+1}^{k_F}$) and the extracted flow vector ($Q_{k_I+1}^{k_F}$):

$$U_{k_I+1}^{k_F} = [u(k_I + 1)^T + \cdots + u(k_F)^T]^T$$
$$Q_{k_I+1}^{k_F} = [q(k_I + 1)^T + \cdots + q(k_F)^T]^T \quad [3.3]$$

We can define the prediction vector as a sum at every term at every time step.

$$X_{k_I+1}^{k_F} = [A_1]x(k_I) + [A_2]U_{k_1}^{k_F} + \cdots + [A_n]Q_{k_1}^{k_F} \quad [3.4]$$

For the numerical analysis, the grid used in the numerical method is divided into a huge number of nodes, where every node coincides with a cross-section. The canal is divided into many cross-sections, but the water level is only measured in some of them, called checkpoints. The values of the water level at checkpoints are contained in the water-level state vector y(k+1) including n_y elements of the state vector x(k+1).

On the other hand, we can compile the water-level state vector for all time instants in the so-called prediction output vector:

$$Y_{k_I+1}^{k_F} = [y(k_I + 1)^T + \cdots + y(k_F)^T]^T \quad [3.5]$$

The $Y_{k_I+1}^{k_F}$ vector contains some of the elements of the prediction vector $X_{k_I+1}^{k_F}$.We write this vector as “Y”.

Now, let us consider a new vector, which will be essential when we introduce the optimization problem involved in our predictive controller. This vector contains the desired values for the prediction output vector (desired water levels at checkpoints) during the predictive horizon. This vector defines the target values in our optimization problem:

$$\overline{Y}_{k_{I+1}}^{k_F} = [\bar{y}(k_I + 1)^T + \cdots + \bar{y}(k_F)^T]^T \quad [3.6]$$

The desired output vector is called $\bar{Y}$.

The most important features of a linear model, which is defined in equation [3.1], are the variables which define the model. The prediction vector during a predictive horizon (k_I+1, k_F) is defined for the state vector at the time step k_I at every cross-section and the control actions implemented during the predictive horizon. The more the cross-sections are defined by the state vector, the more accurate the predictive vector will be compared with the reality. We could come to the same conclusion talking about the prediction output vector.

The matrices [A] and [B] are built by constant elements representing the physical parameters of the canal. This is not completely correct because these parameters represent coefficients, such as the Manning roughness coefficient, which could change over time.

However, the elements of those matrices [A] and [B] can be evaluated offline and these elements can be acquired in certain periods of time.

3.2.2. *Nonlinear model*

The nonlinear models solve the full Saint-Venant equations simultaneously for unsteady flow along the canal. They provide the most accurate solutions available for calculating an outflow hydrograph, while considering the effects of canal storage and wave shape [BED 02]. The models are categorized by their numerical solution schemes, which include the method of characteristics, finite differences, finite element methods, finite volumes and others.

The method of characteristics is commonly used and it is based on the characteristic form of the governing equations. The two partial differential equations (Saint-Venant's equations) are replaced with four total differential equations and solved in the x, t domain. The four equations are commonly solved using explicit or implicit finite difference techniques [AME 66, LIG 67, BAL 68, ELL 70, STR 70, GÓM 88].

Numerical solution schemes are often referred to as being explicit or implicit. When a direct computation of the dependent variables can be made in terms of known quantities, the computation is said to be explicit. Much research has been performed on this topic [GAR 69, LIG 67]. When the dependent variables are defined by coupled sets of equations, and either a matrix or iterative technique is needed to obtain the solution, the numerical method is said to be implicit. Again, this topic has been well-researched by Amein and Chu [AME 75] and by Amein and

Fang [AME 70], among others. The implicit method has fewer stability problems and can use larger time steps than the explicit method.

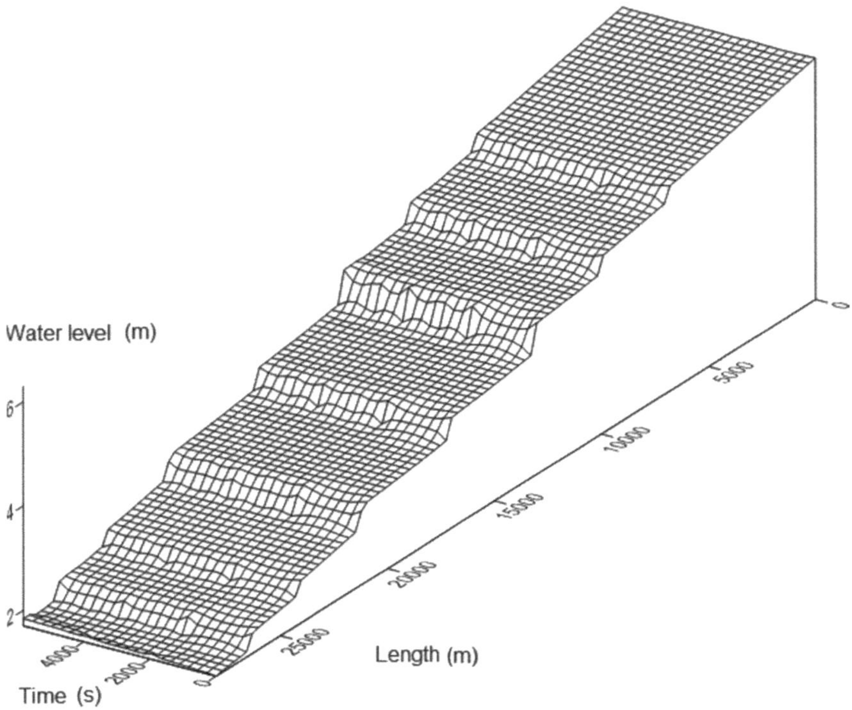

Figure 3.3. *Display of the values obtained by a computer model for a canal with eight pools with a total length of 28 km during a predictive horizon of 6000 seconds. The canal is in steady state; therefore, the gate position, changes in flow rate and demand deliveries are fixed over time. The displayer used in this case is the Surfer 8*

Using nonlinear models in control algorithms is motivated by the possibility of improving the accuracy of the model that represents the system by introducing the nonlinearities of the process. The applicability of nonlinear models in forecasts is a point of discussion which is still open, albeit less so nowadays since current computers solve the full Saint-Venant equations very quickly. In this way, for transport and canal networks where the watermaster wants to keep close to a steady state without huge disturbances, the potential improvement of using nonlinear models may seem to be small. However, in canals where the state vector has a significant variation in water levels and flows, mainly due to external disturbances, the advantages of using nonlinear models can be significant.

We can define a nonlinear model as follows:

$$x(k+1) = f[x(k), s(k+1)]$$
$$s(k+1) = u(k+1) + \cdots + q(k+1) \qquad [3.7]$$

where f is a vector function that depends on the state vector of a previous time step (k), the sluice gate trajectories at time step k+1, the changes in flow rate at time step k+1, the scheduled demands at time step k+1, etc.

The vector function is also dependent on the same physical parameters as in a linear model: the Manning coefficient and local losses coefficients.

We have chosen a nonlinear model that uses the characteristic curves method to solve the full Saint-Venant equations in this book due to the fact that the computation time for solving these equations is not a problem for current computers.

3.3. Optimization problem

3.3.1. *Introduction*

In the simplest case, an optimization problem consists of maximizing or minimizing a function by systematically choosing input values from an allowed set and computing the value of the function. This function is the so-called objective function.

The basic components of an optimization problem are defined as:

– The **objective function**. It expresses the aim of the model, which is either to be minimized or maximized. For example, in a manufacturing process, the aim may be to maximize the profit or minimize the cost. Comparing the data prescribed by a user-defined model with the observed data, the aim is to minimize the total deviation of the predictions based on the model from the observed data. In designing a predictive control for a canal, the goal is to minimize the error between the measured and the desired water level minimizing the gate movements.

– The **control action variables** of the objective function. In the manufacturing problem, the variables may include the amounts of different resources used or the time spent on each activity. In the design of a predictive control for a canal, the variable could be the sluice gate trajectory during a predictive horizon.

– A **set of constraints** are those that allow the unknowns to take on certain values but exclude others. In the manufacturing problem, we cannot spend a negative amount of time on any activity, so one constraint is that the "time" variable cannot be negative. The set of constraints for a predictive control algorithm for canals could be to limit the gate movements between time steps, or the minimum and maximum sluice gate positions.

The objective of an optimization problem is to find values of the variables that minimize or maximize the objective function while satisfying the constraints. In optimization problems, the data prescribed by a model does not match with the target data, and the main goal is to make these values match by modifying a set of unknowns or control variables.

Which would be the objective function in our optimization problem? The objective function is defined by the deviation between the model data (i.e. the prediction output vector) and the desired output vector. On the other hand, the control action variables could be the gate trajectory vector or the extracted flow vector.

We can classify the optimization methods as follows:

– The classic methods are the common methods used in optimization, among them are lineal optimization, non-lineal optimization and dynamic optimization.

– The other optimization methods are the metaheuristic optimization methods intimately linked to artificial intelligence, for instance, evolutionary algorithms or the simulated annealing method [THE 99].

The main difference between both optimization methods is that the classical methods calculate a local minimum and guarantee a local minimum, while the metaheuristic optimization methods calculate a global minimum, but they do not guarantee reaching a global minimum.

The classic methods are robust and the computational cost is very low. These methods have been used in control algorithms giving excellent results. An optimization problem subject to constraints can be stated as follows:

$$To\ find\ X = \begin{pmatrix} x_1 \\ x_2 \\ \vdots \\ x_n \end{pmatrix}\ which\ minimizes\ f(X)\ given \qquad [3.8]$$

$$g_i(X) \leq 0, \quad i = 1,2, \dots, m$$

$$l_j(X) = 0, \quad i = 1,2, \dots, p$$

where X is an n-dimensional vector called the design vector, f(X) is the objective function, and $g_i(X)$ and $l_j(X)$ are the inequality and equality constraints, respectively. When the constraints $g_i(x)$ and $l_j(x)$ are active, the optimization problem is constrained.

The number of variables "n" and the number of constraints "m" and/or "p" need not be related in any way. This type of problem is called a constrained optimization problem.

If the locus of all points satisfying $f(x_i)$ is equal to a constant "C_i" (Figure 3.4), it can form a family of surfaces for several control action variables (x_i) in the design space called the objective function surfaces. When we draw the objective function surfaces with the constraint surfaces (see Figure 3.4), we can identify the optimum point. When we have four or more design variables, we have to solve the problem as a mathematical problem and this visualization is not possible. It is possible graphically only when the number of design variables is three or lower.

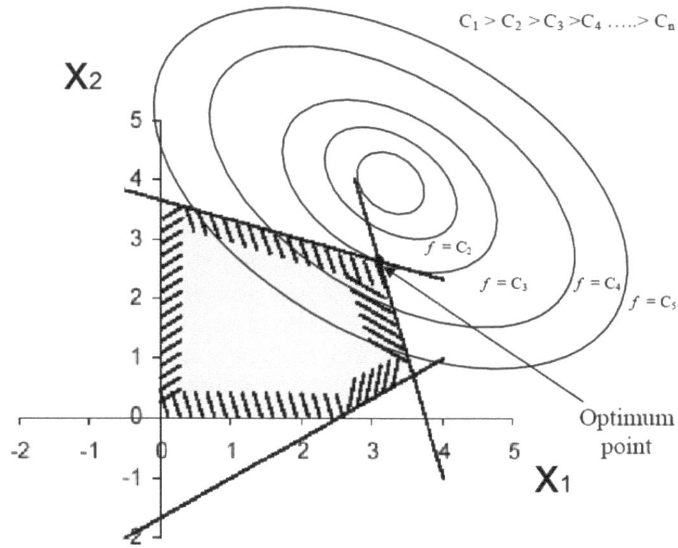

Figure 3.4. *Optimum point in a constraint optimization problem. For a color version of the figure, please see www.iste.co.uk/bonet/canals.zip*

Optimization problems can be defined without any constraints and they are called unconstrained optimization problems. The field of unconstrained optimization is quite large, and there are a lot of algorithms and software available. In this book, we will use both constrained and unconstrained optimization problems.

3.3.2. *Desired output vector*

There are many ways of defining the water offtake along a canal during a given forecasting horizon, for instance, side exits by gravity through orifices (as in Figure 3.5) are probably the most widely used methods, although the pump station is common too. The gravity offtake is very common because small deviations in the water level do not greatly affect the flow rate deliveries. This kind of hydraulic structure is not able to dry up the canal because the orifice is usually set above the canal bottom. By contrast, the pump is able to dry the canal depending on the extracted flow. In order to maintain an operational water offtake, the water level at the target must be higher than y_0 (Figure 3.5), and the flow rate delivered by the orifice is related by the square root of the water level above y_0. The control of the delivered flow rate is kept by controlling the water level.

On the other hand, we could set a certain desired water-level value at every orifice offtake to achieve a particular scheduled delivery during a predictive horizon, as the flow rate depends on the water level at the orifice offtake. The watermaster keeps the desired water level at the checkpoints to fulfill the scheduled supplies and the farmers obtain a hydrograph that meets their needs. Sometimes a farmer ignores the irrigation scheduling and the disturbance caused will affect the scheduled deliveries from other farmers. Although it is impossible that an external disturbance does not affect the scheduled deliveries for others farmers, the feedback controller must recompute a new sluice gate trajectory to quickly reach the desired water level at all checkpoints, restoring the scheduled demand deliveries as soon as possible.

External disturbances are unknown elements in the process, and therefore, they do not provide an easy way to estimate the desired water level. The desired water level is the value required in order to meet the needs of the farmers, and the deviations between the desired water level and the real one are proportional to the deviations between the delivered flow rate and the scheduled delivered flow rate. There is a direct link between the disturbance and the deviations of demand deliveries. The delivery schedule is fulfilled at the same time that the water-level measurements reach the desired water level at particular cross-sections.

The water-level measurements are performed in particular cross-sections or checkpoints usually located next to a sluice gate or pump station. As we will see in the following chapters, the best mathematical approximation for an optimization problem would be to use the water level and velocity at every cross-section in the canal and to calculate the deviation between the desired and measured water levels at each cross-section. However, it would be impossible to get measurements at every cross-section of the canal.

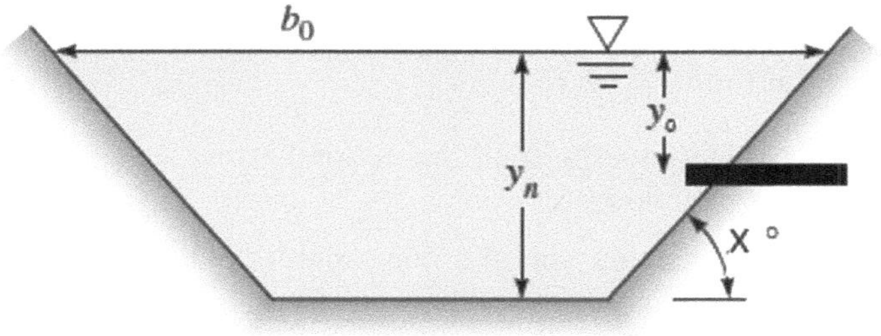

Figure 3.5. *Orifice offtake in a cross-section*

The aim of the objective function is to minimize the error between the measured and the desired water level, and this function may be chosen to meet a variety of conditions, as we will see in the next section.

3.3.3. *Objective function*

The goal of the optimization process is to find the parameter values that result in a maximum or minimum of a function called the objective function. We can classify the objective function as follows:

– the type of objective function: linear or quadratic;

– the variables involved in the objective function;

– the checkpoints or target points evaluated by the objective function.

The objective function defined in equation [3.8] should be minimized and the complexity required to derive this function is linked to the type of objective function and the constraints.

In predictive control in canals, the variable involved in the objective function could be the water level and the variability between the computed and desired water levels at the checkpoint would be the error that should be minimized. This error can be established in the objective function in different forms, quadratic [3.9] or linear.

$$J_{k_1}^{k_F} = \frac{1}{2}\left\| {r'}_{k_1}^{k_F} \right\|_T \quad quadratic\ objective\ function$$

$${r'}_{k_1}^{k_F} = Y_{k_1}^{k_F} - \bar{Y}_{k_1}^{k_F} \qquad \|r'\|_T = r'^T[Q']r' \qquad [3.9]$$

$$T\ is\ a\ weight\ matrix$$

From the type of objective function and constraints, we could also classify the optimization problem as follows:

– if the objective function and the constraints are linear, the optimization problem is linear;

– if the objective function is quadratic and the constraints are linear, the optimization problem is quadratic;

– if the objective function is quadratic and the constraints are nonlinear, the optimization problem is nonlinear.

The variables involved in the objective function could be the water level or the flow rate, and the variability between the computed and desired water levels or the computed and desired flow-rates ($\bar{Q}_{k_{I+1}}^{k_F}$) at the checkpoint is the error that should be minimized. The next equation shows this kind of objective function:

$$J_{k_1}^{k_F} = \frac{1}{2}\left\| {r'}_{k_1}^{k_F} \right\|_T or\ \frac{1}{2}\left\| o_{k_1}^{k_F} \right\|_{T'}$$

$${r'}_{k_1}^{k_F} = Y_{k_1}^{k_F} - \bar{Y}_{k_1}^{k_F};\ \|r'\|_T = r'^T[Q']r' \qquad [3.10]$$

$$o_{k_1}^{k_F} = Q_{k_1}^{k_F} - \bar{Q}_{k_1}^{k_F}\ ;\ \|q\|_{T'} = q^T[Q']q$$

We have introduced the water level and the flow rate, both of which are controlled variables, to define the objective function, but we could use a control action variable such as the gate openings (U) to evaluate the objective function. The main goal is to minimize the deviation between the desired and measured water levels, but it is also important to minimize the sluice gates' movement, because excessive gate movement causes an engine wear of the gate and is a waste of energy for every gate movement, which could otherwise be avoided. The objective function can be dependent on both. The next equation shows this objective function:

$$J_{k_1}^{k_F} = +\frac{1}{2}\left\|r'^{k_F}_{k_1}\right\|_T + \frac{1}{2}\left\|s_{k_1}^{k_F}\right\|_{T'}$$

$$s_{k_1}^{k_F} = U_{k_1}^{k_F} - \overline{U}_{k_1}^{k_F} \quad [3.11]$$

where $\overline{U}_{k_1}^{k_F}$ is the gate trajectory with minimum operating cost.

In predictive control, the gate position could be a variable involved in the objective function too, but we would have to define the gate trajectory with a minimum operating cost, which we call the desired gate trajectory.

On the other hand, it would be excellent to have a large number of checkpoints to establish a more effective control of the canal state, but the checkpoints or target points evaluated by the objective function are usually limited by the number of sensors.

4

Overall Control Diagram

4.1. Introduction

The main objective of this book is to develop an online predictive control able to operate the canal in real time, rectifying the gate trajectory in case of disturbances and establishing the desired behavior of the canal by the watermaster. In our opinion, an on-line predictive controller, operating alone and only using as input data the water levels measured at the checkpoints, is not an accurate predictive controller. There are factors that largely affect the canal as disturbances or changes in the Manning coefficient that the online predictive controller needs to know to recalculate the optimal sluice gate trajectory. For this reason, we define a set of algorithms that operate altogether in parallel; however, not all of them necessarily operate in real time, as the period of time that each one is working depends on the variability of the process in time. All of the information obtained by these algorithms is supplied to the online predictive controller to meets its objective.

4.2. Structure Diagram

Although there are multiple forms to define an overall scheme for canal control, we show our ideal overall control diagram of a canal in Figure 4.1. This figure will be referred to in several chapters in this book, because the algorithms developed follow the structure shown in this diagram. Every process, as well as the task developed for each algorithm, is introduced below (Figure 4.1).

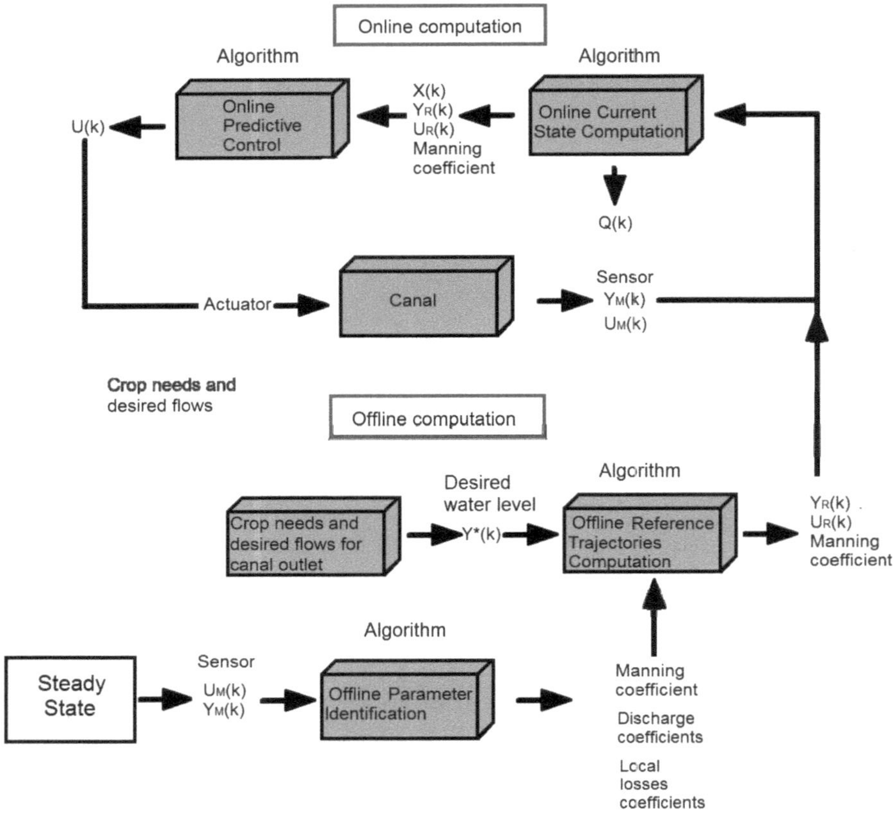

Figure 4.1. *Overall control diagram of irrigation canal*

1) "*Crop needs and desired hydrographs for canal outlets*": The hydrographs at the lateral diversion points of the main canal are calculated on the basis of the water demands. They are fixed considering the requirements of farmers and others, which are accepted by the watermaster. The behavior of the canal supplying these hydrographs determines the "desired behavior" (Y*) at several cross-sections (Figure 4.2).

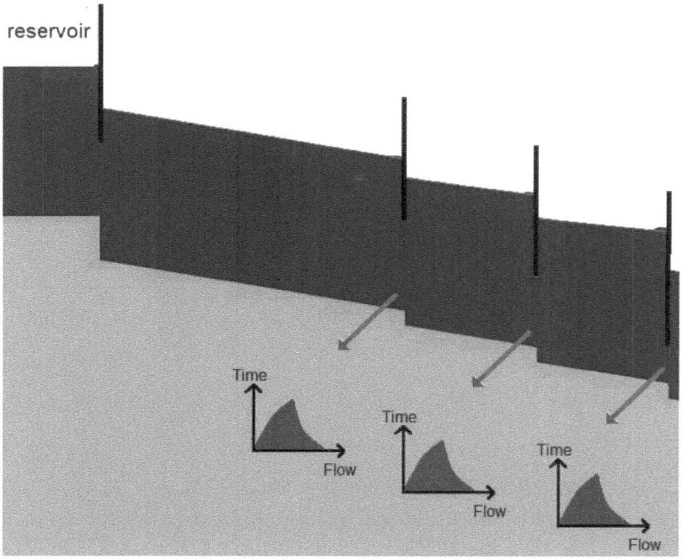

Figure 4.2. *Hydrographs at the lateral diversion points of the main canal*

2) "*Offline Computation of the Reference Trajectories*": The desired behavior (Y*) must be transmitted to the "Reference Trajectories Calculation" algorithm that determines the position of each gate. This algorithm calculates the optimum behavior (Y_R, reference water level), which is the one most similar to the desired behavior that is physically possible. We denote "U_R" the optimum gate trajectories calculated to obtain the optimum behavior (Y_R), which must be calculated off-line (e.g. with an anticipated irrigation cycle). There is extensive literature on feedforward control algorithms that compute the reference trajectories, such as GoRoSo [SOL 03], Bautista *et al.* [BAU 03] and Wahlin *et al.* [WAH 03].

3) "*Offline Parameter Identification*": In case of variations of empirical parameters, such as the Manning coefficient, the water-level measurements obtained during a previous irrigation cycle in a steady state could be transmitted to the "Offline Parameter Identification" algorithm to estimate these empirical parameters (Figure 4.1). The values of these parameters could be sent to the "predictive controller" and the "online current state computation" to be more accurate in their respective processes. This algorithm could also operate online and update the empirical parameters during the irrigation cycle.

4) "*Digital to analog converter – D/A and A/D*": A computer or control system based on a microprocessor cannot interpret analog signals and uses only digital signals, which need to be translated or converted into binary signals.

Figure 4.3. *DAC used to convert digital signal to analog*

A DAC converts an abstract finite-precision number (usually a fixed-point binary number) into a physical quantity (e.g. voltage or intensity) (Figure 4.4). Control actions on the canal are implemented by means of digital–analog transformations (D/A), and the average measurements obtained for each sampling period ($y_M(k)$ and $u_M(k)$) are obtained by means of analog–digital transformations (A/D). The water-level measurements are converted and transmitted to the "Online Current State Computation" algorithm (A/D converter). On the contrary, the gate trajectories from the "online predictive control" are converted and transmitted to the sluice gate actuator (D/A converter).

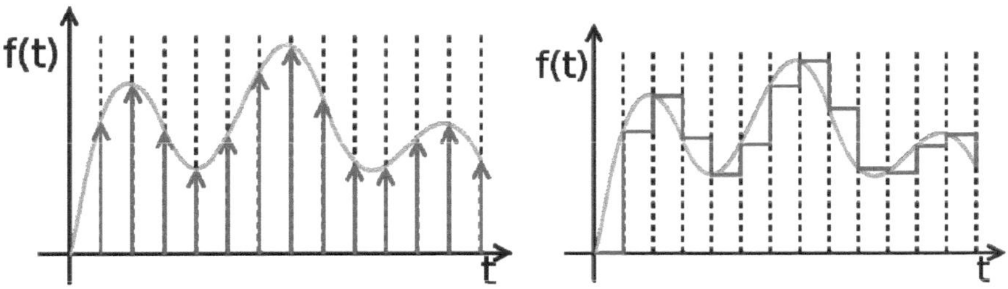

Figure 4.4. *(Left) Ideally sampled signal and (right) piecewise constant output of an idealized DAC lacking a reconstruction filter*

5) "*Online Current State Computation*": This module uses all of the values stored in the database as y_M (k + i) and u_M (k + i), which have been measured at the checkpoints in the canal during a past time horizon (i = 0…−λ). A difference in the water level measured at checkpoints from the desired water level is probably due to a disturbance introduced into the system or erroneous roughness coefficients used in

the calculation. Thus, it is very important to update the Manning coefficients periodically. One of the algorithms developed in this module is able to estimate the disturbances that have made the measured water level at checkpoints different from the desired water level. This algorithm can obtain the current hydrodynamic state of the canal, which is very useful in any case. The hydrodynamic state of a canal is defined as the velocity and water level at each cross-section of the canal. To date, there has been no study conducted on that kind of algorithm.

6) "*Online Predictive Control*": These reference trajectories are transmitted to the next algorithm, called "Predictive Control", which must re-act online in case of deviations between the observed data and the desired behavior at the checkpoints in each regulation period (e.g. every 5 min), due to unknown perturbations. The predictive control recalculates new gate positions (u(k)) to return to the reference behavior. There is extensive literature on feedback control algorithms that recalculate the gate trajectories, such as CARA [MAR 89], Wahlin *et al.* [WAH 06] and Clemmens *et al.* [CLE 04].

Figure 4.1 shows a cycle of processes, where some blocks represent processes in real time and other blocks represent offline processes. For example, for the blocks that represent the crop needs and desired hydrographs for canal outlets, offline computation of the reference trajectories and offline parameter identification are processes that can be solved offline, that is, they are not solved in every sampling period K. The remaining blocks should be solved online, using data collected at every sampling period. The large restriction in real-time problems is the computational time, which must be as short as possible and always lower than the regulation period, because the sluice gate actuator receives new gate trajectories from the control algorithm at every regulation period.

The blocks that calculate the reference trajectories are updated periodically, the frequency of which is determined by the irrigation cycle. The reference trajectories are calculated at every irrigation cycle according to the scheduled deliveries and the changes in the Manning coefficient.

The data recorded from the beginning of the irrigation cycle, such as the measured water level at checkpoints, can be used to estimate the current state of the canal; the output data obtained in solving this problem becomes the input data for the "online predictive control". In this last block (online predictive control), the predictive control recalculates a new sluice gate trajectory if the water level at the checkpoints has changed from the desired water level, which depends on the current canal state at each regulation period. To complete each cycle of the overall control diagram, the control algorithm sends a set of sluice gate positions to the D/A converter, which then sends a signal to the sluice gate actuator.

From a theoretical viewpoint, the input data used in an online predictive control are obtained in different processes; therefore, not all input data used in the online predictive control are obtained at the same time. The overall control diagram contains several processes, and data obtained in each of them are used in different blocks, each of which is defined in a particular step in the overall diagram.

For instance, the offline parameter identification is not a process estimated online, because the main objective is probably to estimate the Manning coefficient, which is not susceptible to change every regulation period. This process is evaluated at the beginning of the irrigation cycle. The offline parameter identification is a quite important part in the overall diagram, because the canal state is very sensitive to changes in the Manning coefficient, even to the point that a good control algorithm would be useless if calibrated with wrong Manning coefficients, as we will show in Chapter 7. In that sense, the ASCE Task Committee on Canal Automation Algorithms developed a series of test cases to test the suitability of canal control, some of which consist in changing the canal conditions, such as the Manning coefficient, to demonstrate the sensitivity of the algorithm to changes in canal conditions over time. In this book, we assume that all physical parameters, such as the Manning coefficient or the energy loss coefficient, are set at the beginning of the irrigation cycle and that they are constant during the irrigation cycle.

Next, we will summarize the rest of the processes involved in the overall diagram.

– The process "*offline computation of reference trajectories*" runs offline. The watermaster knows all scheduled deliveries during the irrigation cycle, and this process calculates the optimal gate trajectories for a predictive horizon, taking into account the scheduled information. This process is estimated by a feedforward controller such as GoRoSo.

– The process "*online current state computation*" runs online because it estimates the canal state at each time step. This process is developed in this book. The algorithm uses data obtained from the beginning of the irrigation cycle to the present time. The data used by this process are the reference water levels, gate trajectories, demand deliveries and the water-level measurements during the predictive past horizon. The solutions of this process are:

- disturbances,

- water level and velocity along the canal at each time period.

– The process "*online predictive control*" runs in real time because it recalculates the sluice gate trajectories if the water levels at the checkpoints have changed from the desired water level values in each regulation period (Figure 4.5). This process is dependent on the online current state process, as the predictive control needs data input about the disturbances in the canal to set the water-level error during the predictive horizon. The input data used by this process are the sluice gate trajectories, pump trajectories, current canal state and demand deliveries. The online predictive control is defined as a feedback controller, and this term is also used to describe closed-loop controllers introduced earlier. The output data of this process are new gate trajectories. This process is developed in this book.

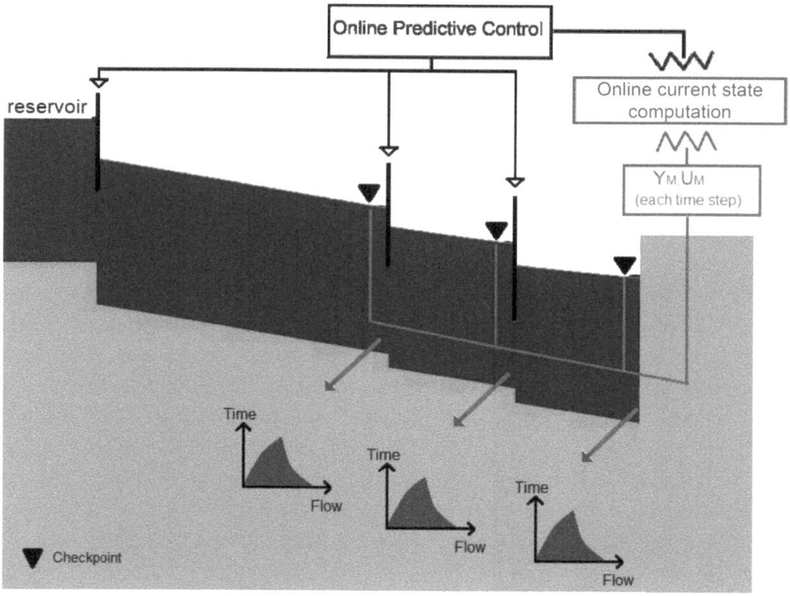

Figure 4.5. *Simple view of an online predictive control on a canal*

4.3. Conclusion

After defining all the process involved in the overall diagram, we have identified different ways in which all of these processes are connected. Are all of these processes easy to include in an overall control diagram? Not at all. For the moment, we will not take into account offline parameter identification because the Manning coefficient can be estimated in advance, as this parameter is almost constant during an irrigation cycle. The offline computation of reference trajectories is not an obstacle, as there are several algorithms in the literature that could calculate the reference gate trajectory in a predictive horizon, such as GoRoSo [SOL 03] or Tomicic [TOM 89]. However, the process which defines the current state of the canal is not the one performed by algorithms found in the literature review, but it is an algorithm introduced in this book, which has not even been proposed before. On the contrary, the process defined as the online predictive controller is the main objective of this book, which needs prediction of the current canal state.

As we will explain in Chapter 5, the first idea was not to design an algorithm to identify disturbances and the current canal state, but to include this algorithm, because the optimal sluice gate trajectory obtained by the online predictive control is dependent on the flow state. The online predictive control needs the more accurate estimation of the disturbances and the hydrodynamic canal state.

If there is not a good prediction of the current canal state, then the gate trajectories solution will be bad because the estimation of the future error will also be bad, although the online predictive controller had been designed from an accurate and robust algorithm.

Some types of controllers, such as the PID, whose algorithm consists of three basic coefficients (proportional, integral and derivative), predict an error based in these three terms. The proportional term produces an output value that is proportional to the current error value; the contribution from the integral term is proportional to both the magnitude of the error and the duration of the error and the derivative of the process error is calculated by determining the slope of the error over time (Figure 4.6). Although this type of controller is able to predict an error for the predictive horizon without knowing the real external disturbance introduced into the system, it would be more accurate in establishing the real disturbance.

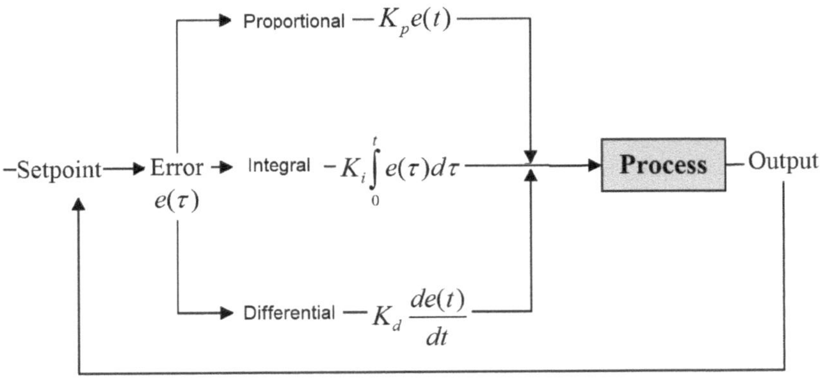

Figure 4.6. *Block diagram of PID controller*

Although the different processes are independent, all of them have in common a similar mathematical approach, that is, they can be formulated as an optimization problem; they use a computer model and they must modify a certain control action variable for setting a measured variable (simulated by the computer model) as close as possible to a particular desired value defined by the watermaster.

5

Canal Survey Information: CSI Algorithm

One of the main problems in the water management of irrigation systems is the control of the equitable distribution of water among different orifice offtakes. The difficulty of managing a canal is partly due to a lack of knowledge of the canal state because the scheduled demand is not fulfilled most of the time. The farmers extract more water than scheduled and hence it is impossible for the watermaster to determine the canal state. This was one of the reasons to develop an algorithm (CSI) that is able to estimate the real extracted flow in certain cross-sections during a past time horizon and the hydrodynamic canal state, that is, the water level and velocity along the irrigation canal during the past time horizon.

Both results obtained with CSI could be valuable for a feedback controller. For instance, there are some controllers that operate with a computer model that does not know the unscheduled deliveries introduced in the canal. If the disturbance is significant, then the real hydrodynamic state of the canal is quite different from the canal state obtained by the computer model, and so the data introduced to the feedback controller are not good enough and the controller eventually fails.

Figure 5.1. *Flow control structures in the irrigation ditch of Manresa*

5.1. Introduction

If we introduce a flow change in a particular section of a canal, then we can estimate the perturbations associated with this flow change from a computer model based on the Saint-Venant equations. CSI does the inverse calculations, that is, it calculates the associated flow change from perturbations (water-level measurements at checkpoints).

The CSI algorithm uses the water-level measurements at certain points of the canal, the scheduled demands and the gate trajectories during a past time horizon to obtain the extracted flow in that time horizon and the hydrodynamic canal state at the current time. This mathematical algorithm is introduced in this chapter.

This algorithm was not our priority at the beginning of this book, and the main objective was to develop a feedback controller. However, several problems relating to the predictive controller forced us to change our viewpoint. These problems were identified when testing our feedback controller, introduced in Chapter 8 (GoRoSoBo). Some of these tests (test cases) were proposed by the ASCE [CLE 98] and the results were not satisfactory. The main reasons for the unsatisfactory results were a lack of knowledge of the canal state and the subsequent difficulties in predicting the future behavior of the canal state. This is the reason why knowledge of the canal state at every moment is very important, as we explored in Chapter 4.

The disturbances that usually modify the canal state are changes in flow deliveries to farmers, which are often unknown. In this sense, the CSI algorithm defines the flow water changes in the canal during a past time horizon (see Figure 5.2).

If a disturbance is introduced to the system, then the free surface of the canal will be different from the water surface obtained without flow changes. In this regard, the smaller the disturbance, the more similar the water surfaces will be. Flow changes can be expressed as a sequence of depths and velocities obtained at certain cross-sections in a canal at discrete times. There is only a set of flow changes to define a particular canal state.

When some hydrodynamic parameters (such as the water level) at target points are measured during a past time horizon, it is possible to determine the flow changes introduced in the canal using the CSI algorithm. This is the objective of this chapter.

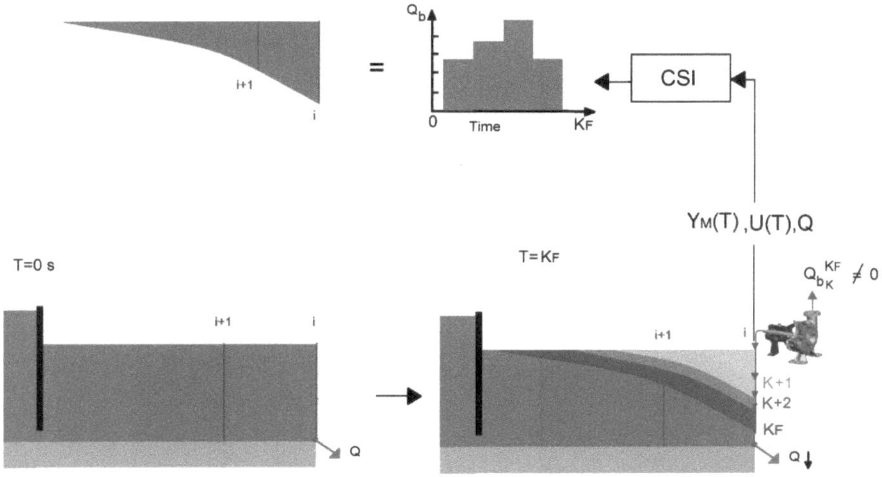

Figure 5.2. *Pumping in a cross-section i. For a color version of the figure, please see www.iste.co.uk/bonet/canals.zip*

In order to achieve this objective, we introduce the HIM (hydraulic influence matrix), which defines the influence of the flow changes on the canal state. The original matrix was defined by Soler [SOL 03], although we have redefined it in this chapter. Soler introduced this matrix in such a way that its purpose is to give us a value of the influence of gate movements on the canal state. The purpose of the matrix is to obtain a value of magnitude of the water level and velocity on the canal due to sluice gate movements. Instead, we introduce a different version of the HIM that establishes the influence of an extracted flow on the water depth and velocity at all points of the canal during a past time horizon. The HIM theory is shown in Appendix 1.

In order to make the meaning of the HIM clearer, we introduce the next example. When we pump in a section of the canal, we modify the canal state, that is, the water level and velocity in particular points at a particular time (Figure 5.3). For instance at t = 10 s, we modify the water level and velocity in cross-sections close to the pump (cross-section i), but we do not modify the hydrodynamic variables in a distant cross-section (i+1). Instead, at t = 40 s, we modify the hydrodynamic variables in all cross-sections from i to i+1. The HIM defines how the extracted flow modifies the hydrodynamic variables in all sections of the canal during a past time horizon, for instance from t = 0 s to t = 40 s in Figure 5.3. In this regard, the HIM evaluates the range of influence of a flow change along the canal during a set period of time.

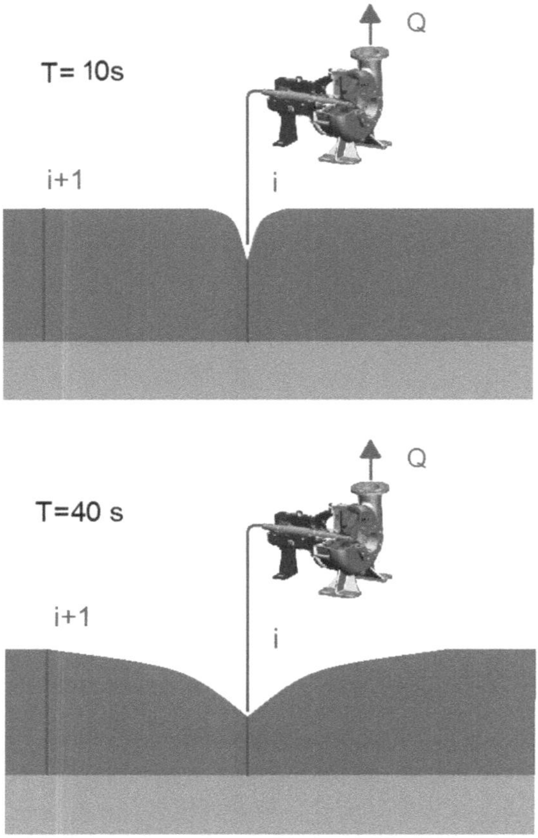

Figure 5.3. *Influence of an extraction flow in the canal state. For a color version of the figure, please see www.iste.co.uk/bonet/canals.zip*

We chose a set of points of the space/time domain of the canal (Figure 5.3), where we could establish the influence of flow change on the water level and velocity at every point (see Figure 5.4). Each element of the HIM establishes the relationship between an extracted flow at a particular section at a particular time step and the velocity/water level of a particular section at a particular time in the canal.

The majority of disturbances that introduce changes in a canal state are caused by flow changes; for instance, the disturbances caused when a farmer holds a pump to get more supplies or when the farmer increases his deliveries from the orifice offtakes are the kind of disturbances considered when we defined the HIM. This matrix gives us the relation between these disturbances and the canal state.

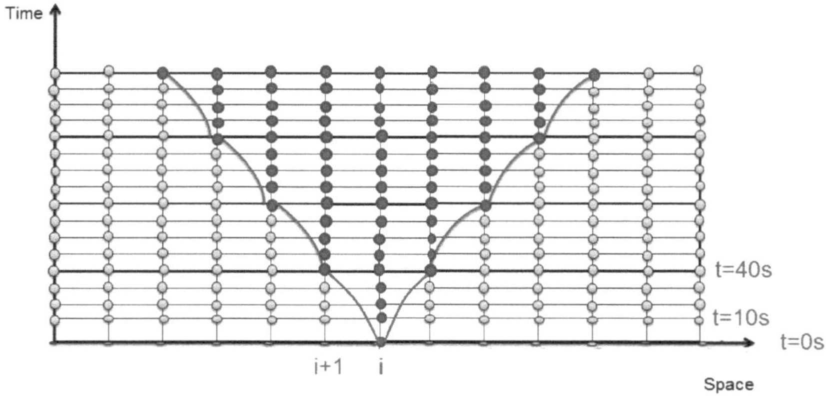

Figure 5.4. *Pumping in a cross-section i of a computational grid. The extracted flow influences the hydrodynamic variables in the section marked with blue dots. For a color version of the figure, please see www.iste.co.uk/bonet/canals.zip*

The hydraulic influence matrix is the main tool in the CSI algorithm, which calculates the history of flow rate changes during a past time horizon in a canal.

5.2. Problem statement

In many cases, the objective of an irrigation canal is to keep the water level constant at certain points where there are orifice offtakes. In this regard, the watermaster fulfills the scheduled demand previously established with the farmers.

Keeping the water level constant is difficult in a canal because there are many outlets along the canal and it is not easy to control the discharges in all of these points. For this reason, we would estimate the unknown flow changes, which modify the desired water level at the checkpoints. More specifically, our objective is to obtain the flow changes (control inputs) over a scenario (past time horizon) in which we know the initial condition of the canal, the gate trajectories, the water level evolution at specific cross-sections (checkpoints) and the offtake scheduling demand. This objective is achieved by the CSI algorithm.

The algorithm establishes an objective function defined from the total deviation between the measured data (water-level measurement) and the data obtained by a computer model (predicted water level), considering an extracted flow. This objective function is minimized from an optimization procedure with the objective to obtain the optimal extracted flow. In this sense, the CSI algorithm solves an inverse problem as an unconstrained optimization procedure.

Some variables and parameters that the CSI algorithm uses to solve the optimization problem are obtained by a computer model describing the free-surface flow in canals, which can be defined from the Saint-Venant equations. These are constituted by a pair of equations that form a quasilinear partial differential system. Moreover, we need to use some other equations to describe the flow along longitudinal structures like siphons and jumps. Also, there are cross-structures for the control of flow, like gates, weirs and offtakes. Because the flow equations do not have analytical solutions, we choose a set of points in the space/time domain to form a computational grid where the solution is computed using a numerical procedure (Figure 5.5).

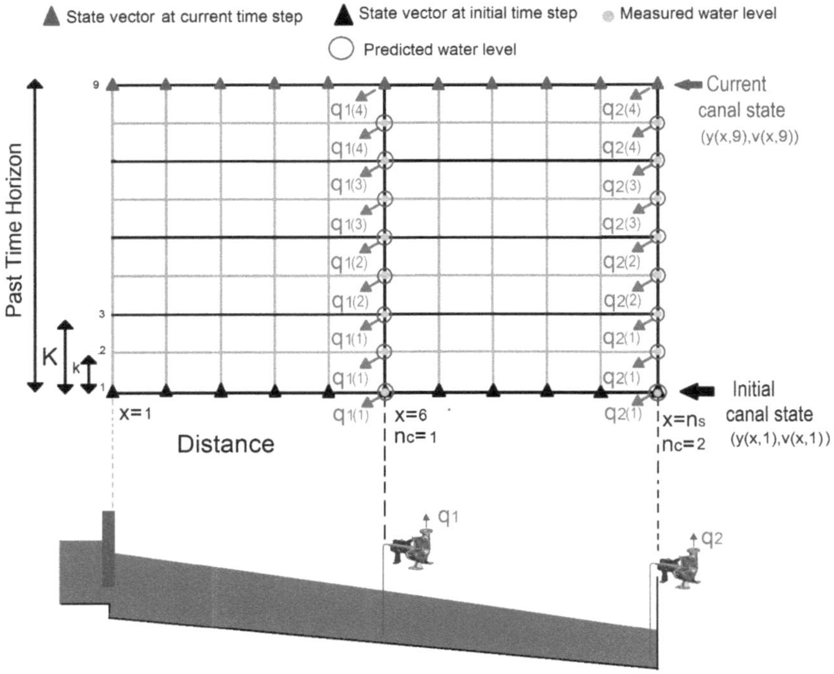

Figure 5.5. *Sketch of a numerical grid of a canal with two pools that is controlled by two checkpoints downstream of each pool. There are pump stations close to each checkpoint. Pump flow trajectories are defined with four operation periods. The figure also shows the x/t-dots where the flow behavior is defined. Note that the uppercase "K" denotes the time interval of control and the lowercase "k" denotes the time instant of simulation. For a color version of the figure, please see www.iste.co.uk/bonet/canals.zip*

5.2.1. *Definition of the optimization problem*

As explained before, the CSI algorithm needs the water-level measurements at certain points (checkpoints) as input data. Now, let us consider a vector, which contains the water-level measurements in the checkpoints at time instant *k*:

$$y_M(k) = \left[y_{m1}(k), \ldots, y_{mi}(k), \ldots, y_{mn_C}(k) \right]^T \qquad [5.1]$$

Finally, all of the vectors in equation [5.1] for each time instant of the past time horizon are joined to define the measured water-level vector, of which the dimension is n_c, which is the number of checkpoints. We define this vector as:

$$Y_M = \left[y_M(1), y_M(2), \ldots, y_M(k_F - 1), y_M(k_F) \right]^T \qquad [5.2]$$

We can check the measured water-level vector values in a computational grid as in Figure 5.5 (yellow dots), of which the dimension is n_y, where $n_y = k_F \times n_c$.

We could obtain the predicted water level from the state vector x(k), which is defined as the vector containing the numerical solution at the time instant k of all of the discretization points:

$$x(k) = \left[y_1(k), v_1(k), \ldots, y_i(k), v_i(k), \ldots, y_{n_S}(k), v_{n_S}(k) \right]^T \qquad [5.3]$$

where $y_i(k)$ and $v_i(k)$ are water depth and mean velocity at point I, respectively, and n_S is the number of cross-sections in which the canal is discretized. In this way, the vector x(0) is the known initial condition.

We can check the state vector at the current time, which defines the current hydrodynamic state in a computational grid as in Figure 5.5 (red triangles) and the state vector at the initial time step (black triangles).

We may include all state vectors (equation [5.3]) for each k-instant of the past time horizon in a single vector which is called the prediction vector:

$$X_1^{k_F} = \left[x(1)^T, x(2)^T, \ldots, x(k_F - 1)^T, x(k_F)^T \right]^T \qquad [5.4]$$

where k_F is the final instant of the past time horizon (Figure 5.5). The dimension of this vector is $n_X = (2 \times n_S) \times k_F$.

Given the initial conditions x(0) and the boundary conditions needed for simulation, the prediction vector is uniquely determined by the extracted flow vector.

Taking into account the fact that in our optimization problem the objective function only evaluates error at particular points, we are only interested in the water level at target points, where we also obtain the water-level measurements. We define a new vector that contains the water depth values given at a prescribed number of points (n_c) for $k=1 \ldots k_F-1$:

$$y(k) = \left[y_1(k), \ldots, y_i(k), \ldots, y_{n_C}(k) \right]^T \qquad [5.5]$$

This vector is constituted by a subset of values of the state vector in equation [5.3]. All vectors shown in equation [5.5] for all of the time instants during the past time horizon are grouped in the so-called prediction output vector. The position of the elements of this vector in the grid domain (Figure 5.5) coincides with the elements of the measured water-level vector:

$$Y_1^{k_F} = \left[y(1)^T, y(2)^T, \ldots, y(k_F - 1)^T, y(k_F)^T \right]^T \qquad [5.6]$$

The dimension of the prediction output vector is $n_Y = k_F \times n_C$. The vector in equation [5.6] contains all water depth values, and it coincides with the yellow dots (*x/t*-dots) shown in Figure 5.5.

The prediction output vector is clearly related to the prediction vector in equation [5.4] in the form:

$$Y_1^{k_F} = C\, X_1^{k_F} \qquad [5.7]$$

where C is a matrix, called the discrete observer matrix by Malaterre [MAL 94], of dimension $n_Y \times n_X$, of which the components are only "zeros" and "ones". This matrix defines the direction of the control logics along a canal pool: downstream-level control, upstream-level control or control of intermediate water levels.

As introduced before, the CSI algorithm calculates the extracted flow at several points (for instance, pump stations) during a past time horizon. In this case, as illustrated in Figure 5.6, the pump stations pump with an operation period K. Then, the extracted flow trajectories are piecewise functions. The extracted flow vector is defined by grouping together all of the extracted flows during the past time horizon, as follows:

$$Q_b = \left[q_1(1), \ldots, q_{n_p}(1), \ldots, \ldots, q_1(K_F), \ldots, q_{n_p}(K_F) \right]^T \quad [5.8]$$

where $n_Q = n_P \times K_F$ is the dimension of this vector, n_P is the number of pump stations and K_F is the number of control operation periods in which the past time horizon is divided.

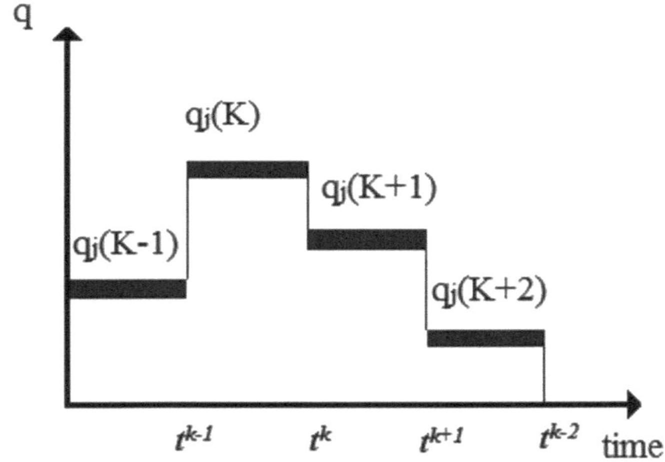

Figure 5.6. *Mathematical representation of a pump flow trajectory*

In this way, only Q_b determines the behavior of the canal along the past time horizon. When the extracted flow trajectories are implemented in the canal, the flow response given by the model will be unique. Inversely, one flow behavior is caused by only one set of extracted flow vectors.

We can check the extracted flow vector values in a computational grid shown in Figure 5.5 (blue arrows).

After estimating the extracted flow vector, the CSI algorithm can also estimate the state vector at the current time, that is, the current hydrodynamic canal state (water level and velocity in all cross-sections of the canal) from the real extracted flow, scheduled demands, the initial conditions and the gate trajectories during the past time horizon.

5.2.2. Optimization problem: objective function

As introduced in Chapter 3, an objective function is an equation to be optimized given certain variables that need to be minimized or maximized using nonlinear programming techniques.

The main objective is to set the flow changes that modify the water level at checkpoints from the measured water levels in the canal. The CSI algorithm calculates the extracted flow trajectories (Q_b), which explain the changes between the measured and predicted water levels.

The objective is to make the prediction output vector more similar to the measured water-level vector by manipulating the extracted flow vector. In mathematical terms, the objective is to obtain the extracted flow vector that minimizes the following performance criterion:

$$J(Q_b) = \frac{1}{2}(Y_1^{K_F}(Q_b) - Y_M)^T[Q'](Y_1^{K_F}(Q_b) - Y_M)$$
$$Minimum\, J(Q_b{}^*) = F[Y_1^{K_F}(Q_b{}^*)] \quad [5.9]$$

where Q' is a weighing matrix with dimensions $n_Y \times n_Y$. This matrix could be used to define the level of importance of the water-level error at a particular checkpoint. We define this matrix as the identity matrix in CSI. The vector Q_b contains the extracted flow trajectories (equation [5.9]).

The minimum of the objective function is also the minimum deviation between the prediction output vector and the measured water-level vector at checkpoints. By minimizing this function (equation [5.9]), we obtain a set of actions (Q_b*) that establish the best possible approximation between measured and computed water levels. This is the goal of the objective function, and Q_b* is the solution (see Figure 5.7).

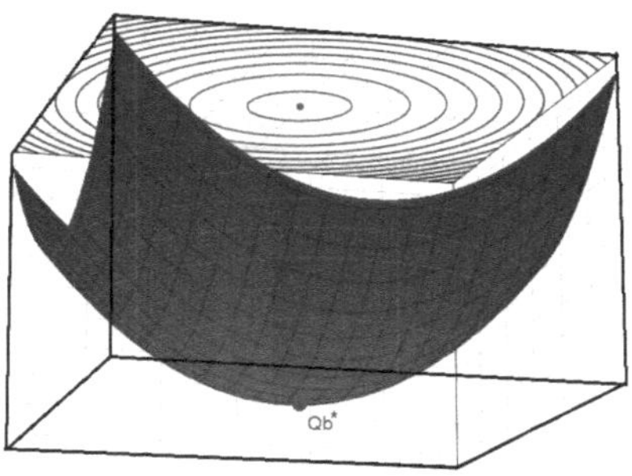

Figure 5.7. *Minimum value (Q_b*) for an objective function. For a color version of the figure, please see www.iste.co.uk/bonet/canals.zip*

5.2.3. *Optimization problem statement*

To summarize this section, the inverse problem is formulated as an unconstrained optimization problem. The optimization problem posed in equation [5.9] is the classical nonlinear problem without constraints. There are many methods to solve an optimization problem in the literature. In this book, we used the Levenberg–Marquardt method to solve this problem, which is introduced in section 5.3. There is extensive literature about unconstrained optimized problems (see [FLE 87] and *Numerical Recipes in Fortran 77* [PRE 97a]).

5.2.4. *Conclusions of the problem statement*

In this book, the input variables for CSI have been directly obtained from a computer code. Therefore, system input variables, such as water level measurements, are obtained from the computer model. As we can know the water-level error at checkpoints, the next step it is to calculate the extracted flow trajectories that better explain the water-level changes at checkpoints. For this purpose, we will use the hydraulic influence matrix (HIM(Q_b)); although the HIM was introduced by Soler [SOL 03], we have modified it as we are interested in obtaining the influence of an extracted flow on the canal state. In this chapter, every element of the HIM represents the influence of an extracted flow at a particular point on the canal state (water level, velocity in a particular section at a particular time step), and so an easy way to find the changes on the canal state when you introduce a change in flow is from the hydraulic influence matrix.

If all flow changes are grouped in an extraction flow trajectories vector ΔQ_b, then, by multiplying the hydraulic influence matrix by this vector, we would obtain the variation of the hydrodynamic variables associated with these flow changes (Figure 5.8) as follows:

$$\Delta X = [HIM(Q_b)]\Delta Q_b \qquad [5.10]$$

where ΔX represents the changes in hydrodynamic variables at checkpoints.

In Appendix 1 at the end of this chapter, we introduce the hydraulic influence matrix, where more details can be found.

The name "hydraulic influence matrix" is derived from the fact that its hydraulic components represent the influence of a flow change over the state vector in all cross-sections at several time steps. The question is, how can we obtain this matrix?

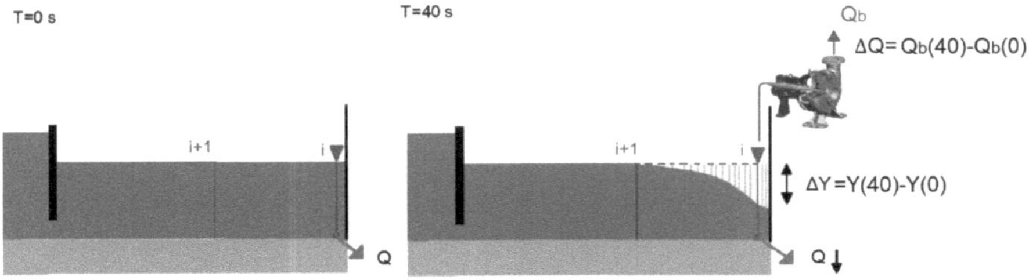

Figure 5.8. *Direct relation between water-level error and extracted flow. For a color version of the figure, please see www.iste.co.uk/bonet/canals.zip*

Based on the full Saint-Venant equations, which describe the free-surface flow in canals, a set of discretized finite difference equations is established to calculate the flow behavior from the extracted flow. The numerical part consists of the application of the numerical method of characteristics to the Saint-Venant equations in order to obtain algebraic equations, and all of the influences are grouped together in a global matrix, which is referred to as HIM(Q_b). Based on this system of equations and using the first derivative $\left(\frac{\partial X}{\partial Q}\right)$ on an analytical process, the changes in flow behavior can be established by a flow change at a point in a certain time instant.

This matrix makes it possible to implement the CSI algorithm as a direct relation between the water level error and flow changes (Figure 5.8). A detailed description of the method to compile the hydraulic influence matrix is presented in Appendix 1.

One of the singularities of this matrix is its condition number, which initially caused numerical problems in CSI. This is an ill-conditioned matrix because its components have varying degrees of influence on the canal state. In this way, it is possible that some rows or columns are close to zero, which results in instabilities in the numerical method. In this case, we had no other choice but to increase the condition number of the matrix as CSI has to inverse the HIM.

5.2.5. *System output/input variables of CSI*

To test the numerical examples, we use a computer model to replace the "measurements". An advantage of using predictive models is that the numerical examples to test are limitless. The disadvantage is that measurements obtained by a predictive model are not as accurate as the measurements taken in a canal, as was introduced in Chapter 3.

The inputs are applied to a control method to predict a control action and influence the output variables of the system. Figure 5.9 shows the input and output variables in the CSI algorithm, taking into account the variables considered in the process, as follows:

– Procedure: CSI algorithm.

– Input variables: scheduled demands, gate trajectories and water-level measurements.

– Output variables: extracted flow vector and state vector at the current time step.

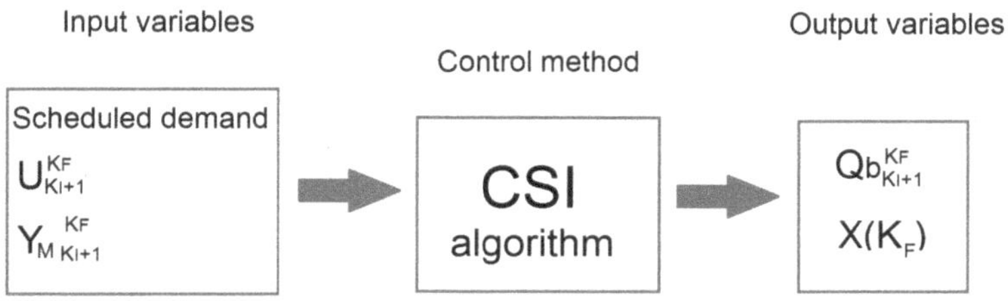

Figure 5.9. *Input/output variables in the CSI algorithm*

5.3. Estimation algorithm from unconstrained optimization

The CSI algorithm solves an unconstrained optimization problem to obtain the extracted flow vector solution, minimizing the objective function. There are several methods to solve an optimization problem, all of which have to reach the solution (Q_b*) and fulfill two conditions to ensure that Q_b* is the better solution. The first condition (equation [5.11]) demonstrates that the solution is a maximum, minimum or saddle, and so the first derivative of function in the solution (Q_b*) is equal to zero. The second condition (equation [5.13]) demonstrates that the solution (Q_b*) is a minimum. These conditions must be achieved by all optimized methods (constrained and unconstrained). In this way, it is interesting to introduce these order conditions.

5.3.1. *First-order and second-order conditions*

There is no auxiliary information mentioned in this section about the proposition and lemmas but there is extensive literature about the conditions in optimized problems (see Fletcher [FLE 87] and Luenberger [LUE 84]).

First-order conditions:

The first-order conditions are necessary to find an optimal solution of an unconstrained optimization problem as follows:

$$\nabla J_{Q_b}(Q_b{}^*) = 0 \qquad [5.11]$$

where $\nabla J_{Q_b}(Q_b{}^*)$ is the first derivative of the objective function around the solution $(Q_b{}^*)$.

We can introduce a definition of a strict local minimizer in which $J(Q_b) > J(Q_b{}^*)$ for all $Q_b \neq Q_b{}^*$ sufficiently close to $Q_b{}^*$. A stronger definition is set for an isolated local minimizer in which Q^* is the only local minimizer in a neighborhood of $Q_b{}^*$.

Therefore, by developing the first-order condition, we obtain:

$$\nabla_{Q_b} J(Q_b{}^*) = (\nabla_{Q_b} X_1^{K_F}(Q_b{}^*)^T - Y^*)^T [C]^T [Q'] ([C] X_1^{K_F}(Q_b{}^*) - Y^*) = 0 \qquad [5.12]$$

The vector $\nabla_{Q_b} J(Q_b{}^*)$ is the first derivative of the objective function. The term $\nabla_{Q_b} X_{K_I+1}^{K_F}(Q_b{}^*)^T$ contains the first derivative of the extracted flow on the state vector. We can express the first derivatives from the water level and velocity variables $(\frac{\partial v_i}{\partial q_{b_i}}, \frac{\partial y_i}{\partial q_{b_i}})$. The dimension of the matrix $\nabla_{Q_b} J(Q_b{}^*)$ is $n_Q \times n_X$.

The gradient of the tangent line must be null over $J(Q_b)$ surface at a stationary point or solution point. Although this is a necessary condition, this is not the only condition, as we will see in the next paragraph.

Second-order conditions:

In fact, all minimization methods are only based on locating a point $Q_b{}^*$ such that $\nabla J_{Q_b}(Q_b{}^*) = 0$. This may not be a strict local minimum and, in general, it is referred to as a stationary point (Figure 5.10). We need a second condition to confirm that $Q_b{}^*$ is a strict local minimum point.

$$s^T \{W(Q_b{}^*, \lambda^*)\} s > 0; \ \forall s \in G(Q_b{}^*) \qquad [5.13]$$

where $G(Q_b{}^*)$ is a set of the feasible directions. In an unconstrained optimization problem, all directions are feasible because there is not any constraint that prevents any direction, and so any "s" is a feasible direction to get $Q_b{}^*$.

$$W(Q^*) = \nabla^2_{Q_b Q_b} J(Q^*) \qquad [5.14]$$

where $W(Q_b{*})$ is the Hessian matrix.

$$\nabla^2_{Q_bQ_b} = \begin{bmatrix} \frac{\partial^2}{\partial q_{b_1}^2} & \cdots & \frac{\partial^2}{\partial q_{b_{n_Q}} \partial q_{b_1}} \\ \vdots & \ddots & \vdots \\ \frac{\partial^2}{\partial q_{b_1} \partial q_{b_{n_Q}}} & \cdots & \frac{\partial^2}{\partial q_{b_{n_Q}}^2} \end{bmatrix} \quad [5.15]$$

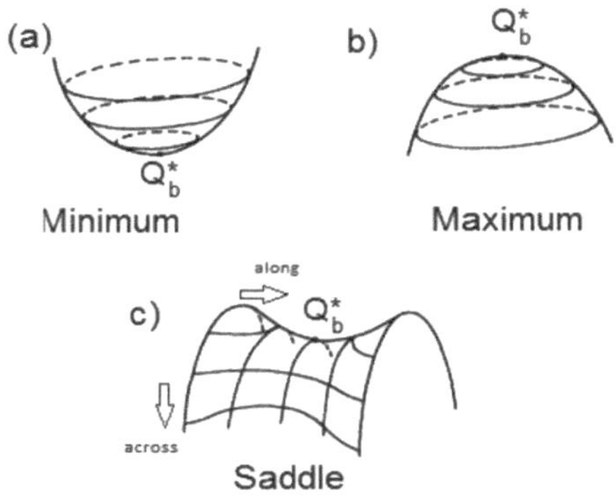

Figure 5.10. *Types of stationary point*

These conditions are necessary and sufficient to demonstrate that $Q_b{*}$ is a strict local minimum.

Equation [5.12] refers to the first-order necessary condition, as it is based on the first-order variations in $J(Q_b{*})$ and therefore the first derivative.

Equation [5.13] is the second-order necessary condition, and it represents the condition that the Hessian matrix is a positive definite matrix. By definition of this property, if the Hessian matrix was an indefinite matrix, then it would be impossible to find a feasible direction ∇Q_b. That is, although $Q_b{*}$ complies with one of the condition requirements (the first condition), we could obtain a value of $J(Q_b{*}+\nabla Q_b)<J(Q_b{*})$, and $Q_b{*}$ would not be a strict local minimum. The last condition could be rewritten as:

$$\Delta Q_b{}^T\left[\nabla^2_{Q_bQ_b} J(Q_b{}^*)\right]\Delta Q_b < 0 \quad [5.16]$$

Meanwhile, all of the graphs have zero slope at Q_b* (Figure 5.10), for there is (a) positive curvature in every direction, (b) negative curvature in every direction and (c) negative curvature across the saddle and positive curvature along the saddle.

As has been written previously, a minimum of the function corresponds to a positive definite Hessian matrix, a maximum to a negative definite matrix and a saddle point to an indefinite Hessian matrix.

We can write the Hessian matrix from the first derivative of the objective function (equation [5.12]) as follows:

$$\nabla^2_{Q_bQ_b} J(Q_b{}^*) = (\nabla_{Q_b} X_1^{K_F}(Q_b{}^*)^T)[C]^T[Q'][C](\nabla_{Q_b} X_1^{K_F}(Q_b{}^*)^T)^T$$
$$+([\mathrm{C}]X_1^{K_F}(Q_b{}^*)^T - Y^*)^T(\nabla^2_{Q_bQ_b} X_1^{K_F}(Q_b{}^*)^T) \quad [5.17]$$

The problem is the second-derivative parameter $(\nabla^2_{Q_bQ_b}\mathrm{X}_1^{\mathrm{K_F}}(\mathrm{Q}_b{}^*)^{\mathrm{T}})$, which is quite difficult to evaluate, but it is a necessary step according to the second-order condition in order to guarantee the solution. To evaluate this second derivative, the algorithm needs a long computational time, and so we use another method to evaluate $\nabla^2_{Q_bQ_b} J(Q_b{}^*)$ without calculating $(\nabla^2_{\mathrm{QQ}}\mathrm{X}_1^{\mathrm{K_F}}(\mathrm{Q}_b{}^*)^{\mathrm{T}})$, taking into account a simplification for the equation.

For Q_b values sufficiently close to Q_b* (i.e. for all Q_b in some neighborhood of Q_b*), the value of $([\mathrm{C}]\mathrm{X}_1^{\mathrm{K_F}}(\mathrm{Q}_b{}^*)^{\mathrm{T}} - \mathrm{Y}^*)^{\mathrm{T}}$ is close to zero. Equation [5.17] can be simplified, omitting the second-derivative term:

$$\nabla^2_{Q_bQ_b} J(Q_b{}^*) \approx (\nabla_{Q_b} X_1^{K_F}(Q_b{}^*)^T)[C]^T[Q'][C](\nabla_{Q_b} X_1^{K_F}(Q_b{}^*)^T)^T \quad [5.18]$$

When $\nabla^2_{Q_bQ_b} J(Q_b{}^*)$ is solved in the simplified form, we use the Gauss–Newton method or least-squares method to solve equation [5.18]. When $\nabla^2_{Q_bQ_b} J(Q_b{}^*)$ is solved in the full form, we use the Newton method to solve equation [5.17].

5.3.2. *Implemented algorithm*

Many techniques for solving this type of problem are available in the literature. The method used to solve the nonlinear optimization problem is the Levenberg–Marquardt method, which is a robust method of easy implementation and a special method to solve ill-conditioned matrices, such as the HIM.

The Levenberg–Marquardt (LM) algorithm is an iterative technique that locates the minimum of a function that is expressed as the sum of squares of nonlinear functions. It has become a standard technique for nonlinear least-squares problems and is widely adopted in a broad spectrum of disciplines. The LM algorithm can be thought of as a combination of the steepest descent method and the Gauss–Newton method. When the current solution is far from the correct one, the algorithm behaves like the steepest descent method: slow, but guaranteed to converge. When the current solution is close to the correct solution, it becomes the Gauss–Newton method. Fletcher [FLE 87], Gill *et al.* [GIL 81] and Luenberger [LUE 84] are useful for solving optimization problems. The theory associated with this method is developed in Appendix 2 at the end of this chapter (section 5.6).

We chose the characteristic curves to describe the temporal discretization of the second-order Saint-Venant equations because it is advantageous over the other methods that we evaluated. This method allows the user to easily understand the physics of wave motion, and the algorithm can use large time steps without significant loss of accuracy, as the Courant–Friedrichs–Lewy condition makes use of much larger possible time steps and consequently saves CPU time.

In the context of simulation, this advantage can be considered as theoretical, although the method uses larger time steps. The solution is also more difficult to find for a large time step than for a short time step because we need to use numerical techniques to solve nonlinear equations. However, if we consider that at each time step the algorithm has to solve a system of equations for each parameter evaluated and build the hydraulic influence matrix, taking into account that each element of this matrix contains a first derivative of the objective function $(\nabla_{Q_b} X_1^{K_F}(Q_b{}^*)^T$, the reduction in time discretization is an advantage.

As was mentioned above, the HIM is an ill-conditioned matrix. In this sense, the LM algorithm gives a certain degree of stability because this method introduces a value in the diagonal of the Hessian matrix to improve the condition of the matrix.

$$H = \left[\nabla^2_{\mathrm{Q}_b\mathrm{Q}_b}\mathrm{J}(\mathrm{Q}_b{}^*)\right] + \nu[I]) \qquad [5.19]$$

where $[I]$ is the identity matrix and ν is the Marquardt coefficient.

The steps defined in the implemented algorithm are shown in the flowchart below, which represents all of the processes carried out by the algorithm. Process operations are represented in boxes, and the arrows define the sequence of operations.

We define the following matrices and variables that are used in the flowchart:

– [Q'] is a weighting matrix.

– [C] is the discrete observer matrix, a matrix of zeros and ones.

– $r' = [C]X_1^{KF}(Q_{b_{it}}) - Y^*$ is the residual vector.

– $X_1^{KF}(Q_{b_{it}})$ is the state vector.

– $[I_M] = I_M[X_1^{KF}(Q_{b_{it}})]$ is the hydraulic influence matrix, all evaluated in $Q_{b_{it}}$.

All of the processes of the algorithm are defined in each iterative loop, where the solution Q_b* is calculated in each iteration (it) for a past time horizon (γ).

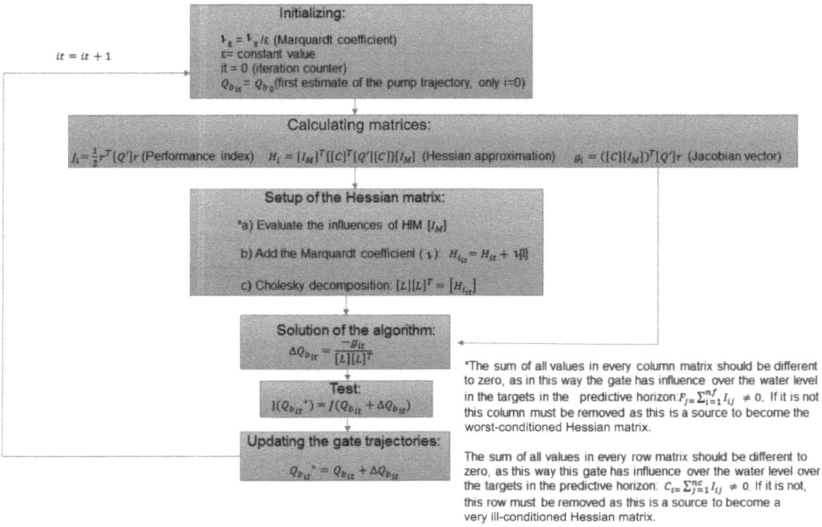

The HIM(Q_{bit}) is an ill-conditioned matrix. In this sense, the matrix obtained by the product matrix of HIM(Q_{bit})^T×HIM(Q_{bit}) is also ill-defined, and for this reason, the matrix determinant is close to zero or negative, and the matrix makes the CSI algorithm unstable. To overcome this problem, we can add to the diagonal matrix H_{it}, the Marquardt coefficient equal to the minimal absolute value of the eigenvalue (H_{it}). In this way, we are sure that the matrix is positive definite and the algorithm is stable.

The choice of the Marquardt coefficient may be made in different ways:

– Selected heuristically and kept constant for a particular canal.

– Selected heuristically and variable at each iteration. The coefficient is divided by a constant value (ε), and so it decreases at each iteration ($\nu = \nu/\varepsilon$).

– Variable at each iteration with a minimal value in order to ensure that the matrix product $HIM(Q_{bit})^T \times HIM(Q_{bit})$ is positive definite. To achieve this objective, we have to obtain the eigenvalues of this matrix and identify the minimal eigenvalue. For obtaining a negative definite matrix, the Marquardt coefficient has to be equal to the absolute value of the minimal eigenvalue. If the matrix is positive definite, the Marquardt coefficient is equal to zero.

We chose the second option; the Marquardt coefficient is variable at each iteration.

5.4. Conclusion

The CSI algorithm calculates the disturbances introduced in a system, and the hydrodynamic canal state during a past time horizon. In this regard, the CSI algorithm is an excellent tool for a watermaster for two reasons. First, the watermaster can know the extracted flow in the canal and the localization of the extraction points. Second, the CSI algorithm is a useful tool for a feedback controller, which needs to know the hydrodynamic canal state in real time.

The CSI algorithm solves an unconstrained optimization problem because it does not need constraints to converge to the solution and the computational time needed to solve an unconstrained optimized problem is shorter than that needed to solve a constrained optimization problem.

The CSI algorithm uses the Hessian matrix in the optimization problem. This matrix is ill-conditioned due to the variability of the influence of the extracted flow on the state vector, because every extraction point has a different value of influence on the checkpoint depending on the distance between them. The influence of a flow change on a checkpoint also depends on the discrete time between the action and the response. To solve the optimization problem, we had to improve the Hessian matrix. This was one of the reasons to use the Levenberg–Marquardt method.

The CSI algorithm solves an unconstrained optimization problem easily, but there are some doubts about the reliability of the algorithm. For instance, the GoRoSoBo algorithm (feedback controller) also initially solved an unconstrained optimization problem, but the solutions obtained by the algorithm were not feasible in some tests, because they were not physically possible. Therefore, the GoRoSoBo algorithm was modified in a constrained optimization problem.

Taking into account the optimization problem solved by the CSI algorithm and the practical examples performed with it, it is clear that the CSI algorithm always calculates a feasible solution. The following points show the importance of the reliability of the algorithm:

– The CSI algorithm calculates an extraction flow trajectory, which is in accordance with the water-level variation between the computed water level and the measured water level at checkpoints. Both water levels are physically possible because the computed water level at checkpoints is the water level at checkpoints without unknown deliveries, whereas the water level measured is directly obtained from measures at checkpoints in the canal. The aim of the CSI algorithm is to find the disturbances caused by unscheduled deliveries. This process is quite different from the process solved by GoRoSoBo, which involves calculating a sluice gate trajectory to reduce the difference between the measured water level and desired water level at checkpoints, and this sluice gate trajectory could either be a feasible solution or not depending on the error between both water levels and the period of time in which the sluice gates have to reach the desired water level.

– The CSI algorithm does not have the same problems associated with the HIM as GoRoSoBo has (without constraints). This feedback algorithm was tested with the test cases proposed by the ASCE [CLE 98], and the gate trajectories obtained by the algorithm in these tests did not have a physical sense. In the test cases, not all sluice gates are close to a checkpoint. In this sense,, the water level obtained at checkpoints allows the algorithm to control the error between the computed water level and the desired water level, but if the sluice gates are close to checkpoints, the excessive sluice gate movements introduced by the algorithm are also controlled. In the test cases, no checkpoint is close to the first gate, which is the most important gate in the canal because this controls the flow between the reservoir and the canal. If the sluice gate only influences the water level at distant checkpoints and the wave generated by the sluice gate movement decreases significantly with the distance, then this sluice gate has a low influence on the water level at the checkpoints. If the algorithm tries to modify the flow conditions using this gate, then it has no other choice but to introduce excessive sluice gate movements to the system. In these specific cases, the optimization problem would require constraints. This is not a situation that affects the CSI algorithm because the extraction points are close to a checkpoint in these tests and the difference between the measured and computed water level is due to specific flow changes.

As we will explain in the following chapters, the CSI algorithm was not the main objective of this book, until we realized the importance of defining the current state of a canal for a feedback controller. The results of the CSI algorithm were excellent, as we will show in Chapters 6 and 7.

5.5. Appendix 1

HIM (hydraulic influence matrix)

Here, we introduce the HIM. We also introduce the equations that define the hydraulic influence matrix and the influence concept of every parameter of the matrix.

The hydraulic influence matrix concept was introduced by Soler [SOL 03], but the theory of this previous work was modified as previously. The CSI algorithm needs the influence of a state variable on the canal state, which is the flow change, for instance, a pump flow. A summary of the HIM, introduced by Soler [SOL 03], is included in his book.

5.5.1. *Free-surface flow equations*

The Saint-Venant equations describe the free-surface flow in prismatic canals and are the result of the application of the principles of mass conservation and of the quantity of movement in a controlled volume of short length in the direction of flow. A rigorous deduction of these equations for prismatic canals can be found in Walker [WAL 87], resulting in:

$$\left.\begin{aligned} \frac{\partial y}{\partial t}+v\frac{\partial y}{\partial x}+\frac{A(y)}{T(y)}\frac{\partial v}{\partial x}=0 \\ \frac{\partial v}{\partial t}+v\frac{\partial v}{\partial x}+g\frac{\partial y}{\partial x}=g\left[S_0-S_f(y,v)\right] \end{aligned}\right\} \qquad [5.20]$$

where x and t are the independent space and time variables, respectively; y is the level of the free surface from the canal bottom; v is the average velocity of all particles of a cross-section of the flow; A(y) is the area of the wet section, which depends on the water level; T(y) is the maximum width, also dependent on the water level; S_0 is the canal bottom slope and S_f (y, v) is the slope friction.

Equations [5.20] will be applied in real time under the following assumptions:

– The curvature of the free surface is small.

– The vertical accelerations are disregarded.

– The pressure distribution along an axis vertical to the liquid is the same as under hydrostatic conditions.

– The canal slope is assumed to be sufficiently small such that its sine is practically null.

– The energy dissipation term is to be specified through the Manning equation, which is used in stationary regime.

$$S_f(y,v) = n^2 \frac{v|v|}{R_H^{\frac{4}{3}}} \qquad \text{with} \quad R_H = \frac{A(y)}{P(y)} \qquad [5.21]$$

where n is the Manning coefficient and P(y) is the wet perimeter, which is a function of the water level, and the changes in flow conditions are not fast enough to generate significant wave fronts.

These equations cannot be solved analytically, only numerically. Thus, a variety of numerical methods exist, and can be found, among others, in Gómez [GÓM 88]. According to Wylie [WYL 69], all of the numerical methods of resolution, explicit, implicit or characteristic, present results similar to reality, depending more on the accuracy of the starting data than on the particular methodologies. Considering this key fact, we use the method of characteristics as it aids physical comprehension of the underlying wave phenomenon in free-surface flow.

Usually, equations [5.20] are expressed in the classical space and time (x/t) axes, but the so-called characteristic curves are used, denoted parametrically by $x^+(t)$ and $x^-(t)$, which locally satisfy the two following differential equations:

$$\left.\begin{aligned} \frac{dx^+}{dt} &= v + c(y) \\ \frac{dx^-}{dt} &= v - c(y) \end{aligned}\right\} \qquad [5.22]$$

Thus, the system [5.20] is transformed into the following two ordinary differential equations:

$$\left.\begin{aligned}\frac{dv}{dt}+\frac{g}{c(y)}\frac{dy}{dt}&=g\left[S_0-S_f(y,v)\right]\\ \frac{dv}{dt}-\frac{g}{c(y)}\frac{dy}{dt}&=g\left[S_0-S_f(y,v)\right]\end{aligned}\right\} \qquad [5.23]$$

where the first is valid along the curve $x^+(t)$ and the second along $x^-(t)$, and c(y) is the wave celerity.

$$c(y)=\sqrt{\frac{gA(y)}{T(y)}} \qquad [5.24]$$

The difficulty of the method lies in the fact that equations [5.22] have to be solved along the characteristic curves or the local axes that are the solution of equations [5.23]. As the latter is a set of nonlinear equations, it obliges us to solve the four equations simultaneously. Fortunately, the curves $x^+(t)$ and $x^-(t)$ always intersect, although they are not orthogonal, and therefore assure hyperbolicity.

In short, solving the system of the two partial derivative equations [5.20] is the same as solving the following set of four ordinary differential equations:

$$\left.\begin{aligned}&(a) \quad \frac{dv}{dt}+\frac{g}{c(y)}\frac{dy}{dt}=g\left[S_0-S_f(y,v)\right]\\ &(b) \quad \frac{dx^+}{dt}=v+c(y)\\ &(c) \quad \frac{dv}{dt}-\frac{g}{c(y)}\frac{dy}{dt}=g\left[S_0-S_f(y,v)\right]\\ &(d) \quad \frac{dx^-}{dt}=v-c(y)\end{aligned}\right\} \qquad [5.25]$$

As the mathematical processes of transformation of the system [5.20] into the equivalent [5.25] are found in many works, such as Gómez [GÓM 88], Soler [SOL 96] and Ames [AME 77], they will not be discussed further here.

The system [5.25] describes the flow conditions in a canal in the same way as the set of equations [5.20] as it adds no new hypotheses to the transformation. However, the system [5.25] is limited in the way in which it is applied. The variable x, which was initially independent, is now dependent on time t, as is understood in [5.22]; then, equation ([5.25] (a)) will be true only along the curves that satisfy equation ([5.25] (b)) and, in the same way, equation ([5.25] (c)) will be true along the solution curves of equation ([5.25] (d)).

The system [5.25] can be represented in the graph x/t, as in Figure 5.11, where the four equations are satisfied at the point of intersection R and therefore the four unknown variables x, t, y and v can be found theoretically. In this way, if the flow conditions at points P and Q are known, then the position of point R can be found and integrated numerically, along with the flow conditions.

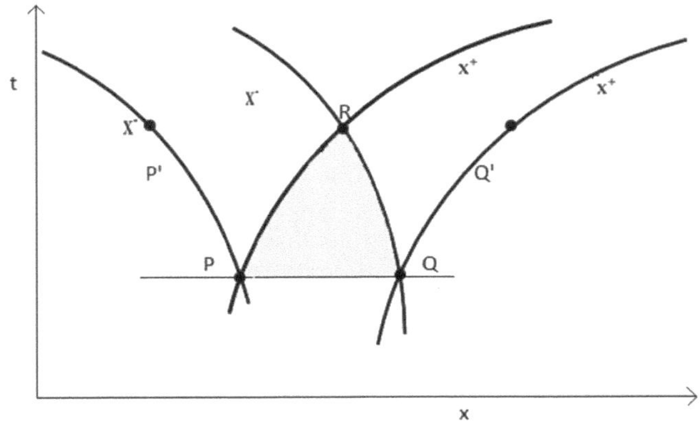

Figure 5.11. *Dependence domain of point R. For a color version of the figure, please see www.iste.co.uk/bonet/canals.zip*

This can be proved thanks to the first of the uniqueness theorems developed by Crandall [CRA 56], in which he showed that if the conditions of flow y and v are known on a curve on the graph x/t, which is not a characteristic curve, such as the line PQ in Figure 5.11, then the set of equations [5.25] determines its solution with uniqueness in the zone marked PQR. This zone is called the dependence domain of point R because the solution at this point is determined exclusively by the conditions of flow produced at any point of this domain. This means that any disturbance introduced at any point of the dependence domain will affect the position of point R and the flow conditions.

A concept that is complementary to the dependence domain is the influence domain. In Figure 5.12, we show the influence domain of point P, that is, the set of points of the graph x/t (the area shaded with red vertical lines) that seem to be affected by the present conditions of flow (y_P,v_P) at this point. In the same way, the area shaded with yellow horizontal lines is the influence domain of point Q and the area shaded with both vertical and horizontal lines is the influence domain of point R.

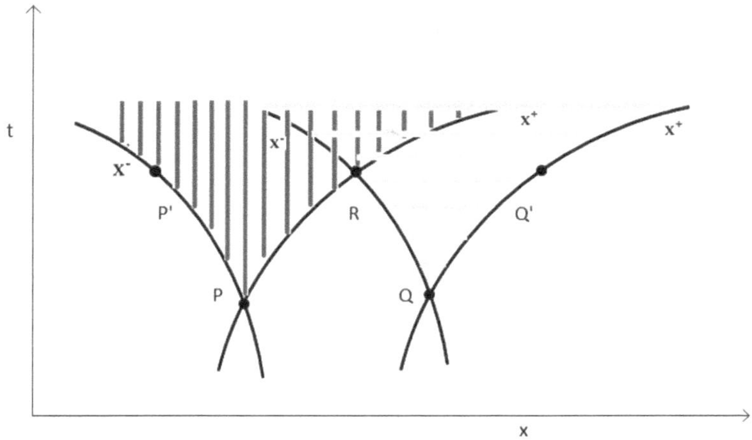

Figure 5.12. *Influence domains of points P, Q and R. For a color version of the figure, please see www.iste.co.uk/bonet/canals.zip*

Once the concepts of influence and dependence domains have been introduced, we can present the basic objective of the following section. When a flow change is made by a pump, a series of changes to the flow condition are produced. First, near to the pump itself and then further away. If the flow change is considered as a disturbance, then we can speak of the disturbance influence domain.

The objective of the next section is to present a way of calculating and quantifying the variation in the conditions of point R, which we will denote Δy_R and Δv_R, when disturbances are introduced to points P and Q, which we will denote Δy_P and Δv_P, and Δy_Q and Δv_Q, respectively.

5.5.2. *Discretization of the characteristic equations*

As previously mentioned, the system [5.20] and the equivalent [5.25] have no known analytical solution, and so the use of numerical techniques has, until now, been compulsory. There are many numerical methods that can be used, at least in a theoretical manner. In this book, we have preferred to use a specific discretization and make the appropriate mathematical developments based on the result of this discretization. In order to have the longest possible integration time period without loss of precision, we have adopted a discretization in finite differences of second order, described as "the method of characteristic curves" in Gómez [GÓM 88]. If this method is applied to equations [5.25] and we take into account the characteristics curves that contain the points P-R and Q-R (Figure 5.12), then we will obtain the following equations:

$$\left.\begin{array}{l}\dfrac{v_R - v_P}{t_R - t_P} + \left[\theta \dfrac{g}{c_R} + (1-\theta)\dfrac{g}{c_P}\right]\dfrac{y_R - y_P}{t_R - t_P} = gS_0 - \left[\theta S_{f_R} + (1-\theta) S_{f_P}\right] \\ \dfrac{x_R - x_P}{t_R - t_P} = \theta\left[v_R + c_R\right] + (1-\theta)\left[v_P + c_P\right] \\ \dfrac{v_R - v_Q}{t_R - t_Q} - \left[\theta \dfrac{g}{c_R} + (1-\theta)\dfrac{g}{c_Q}\right]\dfrac{y_R - y_Q}{t_R - t_Q} = gS_0 - \left[\theta S_{f_R} + (1-\theta) S_{f_Q}\right] \\ \dfrac{x_R - x_Q}{t_R - t_Q} = \theta\left[v_R - c_R\right] + (1-\theta)\left[v_Q - c_Q\right]\end{array}\right\} \quad [5.26]$$

where $S_{fR} = S_f(y_R,v_R)$, $S_{fP} = S_f(y_P,v_P)$ and $S_{fQ} = S_f(y_Q,v_Q)$, and $0 \le \theta \le 1$ is the coefficient of average time that indicates the type of numerical scheme used. When $\theta = 1$, the numerical scheme is implicit; when $\theta = 0$, it is explicit and when $\theta = 1/2$, the numerical scheme is called a central difference or characteristic curve.

If the flow conditions at points P and Q are known, then from the four equations of set [5.26], x_P, t_P, y_P, v_P and x_Q, t_Q, y_Q, v_Q are also known and x_R, t_R, y_R, v_R remain as unknown variables, which can be found by using any of the methods of solving nonlinear equations, such as the Newton–Raphson method. The four equations of set [5.26] can therefore be rewritten as:

$$\left.\begin{array}{l} f_1 \equiv x_R - x_P - \dfrac{1}{2}(t_R - t_P)\left[v_R + c_R + v_P + c_P\right] = 0 \\ f_2 \equiv (v_R - v_P) + \dfrac{g}{2}\dfrac{c_R + c_P}{c_R c_P}(y_R - y_P) - g(t_R - t_P)\left(\dfrac{S_{f_R} + S_{f_P}}{2} - S_0\right) = 0 \\ f_3 \equiv (v_R - v_Q) - \dfrac{g}{2}\dfrac{c_R + c_Q}{c_R c_Q}(y_R - y_Q) - g(t_R - t_Q)\left(\dfrac{S_{f_R} + S_{f_Q}}{2} - S_0\right) = 0 \\ f_4 \equiv x_R - x_Q - \dfrac{1}{2}(t_R - t_Q)\left[v_R - c_R + v_Q - c_Q\right] = 0 \end{array}\right\} \quad [5.27]$$

.

Once the system [5.27] has been solved, we may ask the following question: what would be the solution if, instead of the conditions (y_P,v_P) at point P, we had (y_P+Δy_P,v_P), that is, when a change in water level is introduced? As this flow change perturbs the flow conditions of R, we can linearize to the proximity of y_P and expand a first-order Taylor approximation series as:

$$y_R\left(y_P+\Delta y_P\right)=y_R\left(y_P\right)+\frac{\partial y_R}{\partial y_P}\Delta y_P+O\left(\Delta y_P^2\right) \qquad [5.28]$$

To answer this question, we introduce the concept of hydraulic influence.

We define the hydraulic influence of the conditions of water level and velocity at a point at a given time instant (e.g. point P in Figure 5.11 or Figure 5.12) over the conditions at another point at another time instant (e.g. point R in the same figures), as the disturbance effect that is produced at point R caused by a small modification to the conditions of point P.

In order to answer the given question, suppose that all of the variables of system [5.27] depend implicitly on y_P, that is, y_R (y_P), $v_R(y_P)$, $t_R(y_P)$ and $x_R(y_P)$, and also consider that the implicit function theorem is applied. Then, by solving the 4×4 system:

$$\begin{bmatrix} \frac{\partial f_1}{\partial x_R} & \frac{\partial f_1}{\partial y_R} & \frac{\partial f_1}{\partial v_R} & \frac{\partial f_1}{\partial t_R} \\ \frac{\partial f_2}{\partial x_R} & \frac{\partial f_2}{\partial y_R} & \frac{\partial f_2}{\partial v_R} & \frac{\partial f_2}{\partial t_R} \\ \frac{\partial f_3}{\partial x_R} & \frac{\partial f_3}{\partial y_R} & \frac{\partial f_3}{\partial v_R} & \frac{\partial f_3}{\partial t_R} \\ \frac{\partial f_4}{\partial x_R} & \frac{\partial f_4}{\partial y_R} & \frac{\partial f_4}{\partial v_R} & \frac{\partial f_4}{\partial t_R} \end{bmatrix} \begin{pmatrix} \frac{\partial x_R}{\partial y_P} \\ \frac{\partial y_R}{\partial y_P} \\ \frac{\partial v_R}{\partial y_P} \\ \frac{\partial t_R}{\partial y_P} \end{pmatrix} = -\begin{pmatrix} \frac{\partial f_1}{\partial y_P} \\ \frac{\partial f_2}{\partial y_P} \\ 0 \\ 0 \end{pmatrix} \qquad [5.29]$$

the values $\frac{\partial x_R}{\partial y_P}\frac{\partial y_R}{\partial y_P}\frac{\partial v_R}{\partial y_P}$ and $\frac{\partial t_R}{\partial y_P}$ can be found.

It should be noted that, as equation [5.29] is the result of the application of the implicit function theorem on, equations [5.27], it needs to fulfill the condition that the matrix in equation [5.29] can be inverted. That is, its determinant is non-zero.

It can be shown that this condition is fulfilled and therefore the values of the implicit derivatives $\frac{\partial x_R}{\partial y_P}\frac{\partial y_R}{\partial y_P}\frac{\partial v_R}{\partial y_P}$ and $\frac{\partial t_R}{\partial y_P}$ always exist and are unique. The matrix structure in equation [5.29] can be formed for the remaining "disturbable" variables v_P, y_Q and v_Q. In general, and for future developments, we will refer to a generic variable ψ to denote variables of flow description such as depth, velocity, physical coefficients, gate position and pump flow. Thus, the following general expression will be used:

$$\mathbf{M}\frac{\partial}{\partial\psi}\begin{pmatrix} x_R \\ y_R \\ v_R \\ t_R \end{pmatrix} = -\mathbf{N}\frac{\partial}{\partial\psi}\begin{pmatrix} y_P \\ v_P \\ y_Q \\ v_Q \end{pmatrix} \qquad [5.30]$$

where:

$$\mathbf{M} = \frac{\partial(f_1, f_2, f_3, f_4)}{\partial(x_P, y_R, v_R, t_R)};\quad \mathbf{N} = \frac{\partial(f_1, f_2, f_3, f_4)}{\partial(y_P, v_P, y_Q, v_Q)}$$

We note that equation [5.29] is equal to equation [5.30] when $\psi = y_P$ and thus it has a more general character. From the physical viewpoint, equation [5.30] "moves" or "modifies" the influence of a variable ψ from points P and Q to point R.

The method of calculating the influences shown in this section is closely linked to the numerical scheme of characteristic curves. However, usually this scheme is not exactly used because it gives the solution at a point R whose coordinates (x_R,t_R) are unknown *a priori*. These coordinates are part of the solution and normally it is more important to know the solution of the flow conditions at specific points and at specific time instants. There are two possible ways to solve this problem: (1) first solve and then interpolate, and (2) first interpolate and then solve. The second option is used in this book to solve the problem and is presented in the following section.

5.5.3. *Application to a structured grid*

Figure 5.13 shows how, by placing the characteristic curves net (Figure 5.13(a)) on top of a structured mesh (Figure 5.13(b)), a new scheme is obtained, where the variables for points P and Q are interpolated (Figure 5.13(c)). In this way, we can obtain the flow conditions for the fixed point R.

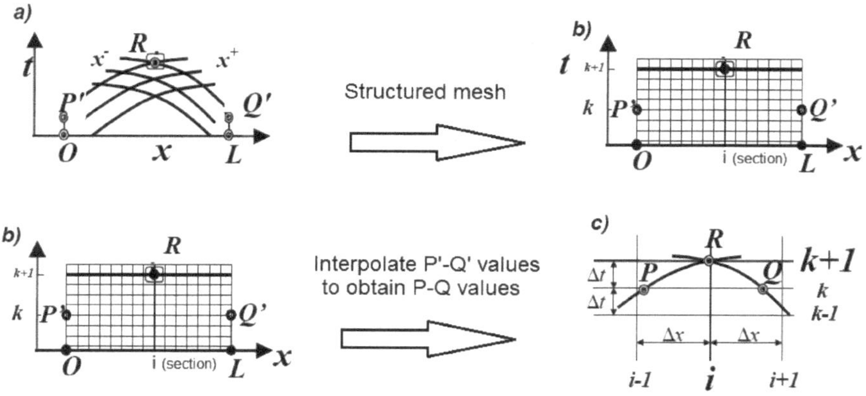

Figure 5.13. *Steps for interpolation onto a structured grid*

Obviously, the same system [5.27] is solved, but now with the new unknowns x_P, y_R, v_R and x_Q. A structured grid like this one creates a new nomenclature. Indeed, each variable will have a double index, where *k* refers to time and *i* represents space. Therefore, y_{ik} and v_{ik} represent the values for water level and average velocity at coordinates $x_i=i\Delta x$ and $t^k=k\Delta t$, where Δx and Δt are appropriately selected. As interpolation also has to be of second order (in order to be coherent with the numerical procedure used), we have used the Lagrange factors (a way of representing quadratic splines). For a dummy variable z, the result is (Figure 5.14):

$$z(x) = s^k\left(x, z_{i-1}^k, z_i^k, z_{i+1}^k\right) = \\ = \left(\frac{x-x_i}{\Delta x}\right)\left(\frac{x-x_{i-1}}{2\Delta x}\right)z_{i+1}^k + \left(\frac{x-x_{i-1}}{\Delta x}\right)\left(\frac{x-x_{i+1}}{-\Delta x}\right)z_i^k + \left(\frac{x-x_i}{-\Delta x}\right)\left(\frac{x-x_{i+1}}{-2\Delta x}\right)z_{i-1}^k \quad [5.31]$$

In this way, variables y_P, v_P, y_Q and v_Q become functions of x_P and x_Q as follows:

$$y_P\left(x_P\right) = s^k\left(x_P, y_{i-1}^k, y_i^k, y_{i+1}^k\right); \quad v_P\left(x_P\right) = s^k\left(x_P, v_{i-1}^k, v_i^k, v_{i+1}^k\right) \\ y_Q\left(x_Q\right) = s^k\left(x_Q, y_{i-1}^k, y_i^k, y_{i+1}^k\right); \quad v_Q\left(x_Q\right) = s^k\left(x_Q, v_{i-1}^k, v_i^k, v_{i+1}^k\right) \quad [5.32]$$

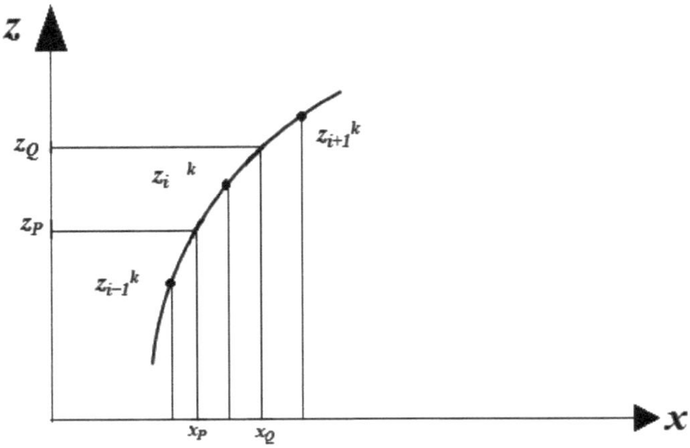

$$z(x)=s(x, z_{i-1}^k, z_i^k, z_{i+1}^k)$$

Figure 5.14. *Interpolation functions*

From now on, it is necessary to determine the system to be solved in order to find the evolution of the influence of a parameter through a structured grid, as has been done in order to find [5.30] from the characteristic curves. Contrary to what might be expected, finding the "transfer" of the influences (or evolution of the influence in time or space) in an interpolation grid simplifies the problem. The reason for this is that the influence of the general parameter on the position and at time instant $\frac{\partial x_R}{\partial \psi} = \frac{\partial x_i}{\partial \psi}, \frac{\partial t_R}{\partial \psi} = \frac{\partial t^{k+1}}{\partial \psi}$ loses all meaning, because we want to find the solution to specific x- and y-axes (Figure 5.15).

If we introduce two different disturbances at time K, then by solving the four equations [5.27] with x_P, $_{yi}^{k+1}$, v_i^{k+1} and x_Q as unknowns, two systems of characteristics, x^+, x'^+ and x^-, x'^-, two solutions for point R, $({}_{yi}^{k+1}, v_i^{k+1})$ and $(y_i^{k+1}, v_i^{k+1})'$, and two sets of interpolated x-axes, x_P, x_Q and $x_{P'}$, x_Q, are obtained, whereas the position of point R remains unaltered.

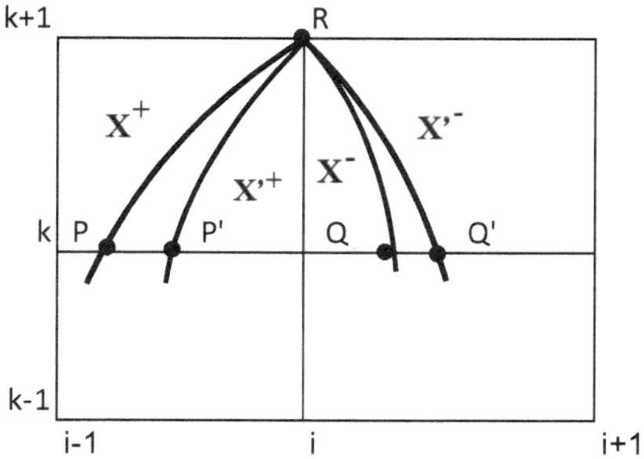

Figure 5.15. *Pair of systems of characteristic curves passing through point R*

Applying once more the implicit function theorem to equations [5.27] with the assumption that $y_{i-1}^k, v_{i-1}^k, y_i^k, v_i^k, y_{i+1}^k, v_{i+1}^k, y_{i+1}^{k+1}$ and v_{i+1}^{k+1} now depend on a general parameter ψ gives a system similar to equation [5.30]:

$$\mathbf{M}\frac{\partial}{\partial\psi}\begin{pmatrix} x_P \\ y_i^{k+1} \\ v_i^{k+1} \\ x_Q \end{pmatrix} = -\mathbf{NS}\frac{\partial}{\partial\psi}\begin{pmatrix} y_{i-1}^k \\ v_{i-1}^k \\ y_i^k \\ v_i^k \\ y_{i+1}^k \\ v_{i+1}^k \end{pmatrix} \qquad [5.33]$$

where:

$$\mathbf{M} = \frac{\partial(f_1, f_2, f_3, f_4)}{\partial\left(x_P, y_i^{k+1}, v_i^{k+1}, x_Q\right)}; \quad \mathbf{N} = \frac{\partial(f_1, f_2, f_3, f_4)}{\partial\left(y_P, v_P, y_Q, v_Q\right)}$$

$$\mathbf{S} = \frac{\partial\left(y_P, v_P, y_Q, v_Q\right)}{\partial\left(y_{i-1}^k, v_{i-1}^k, y_i^k, v_i^k, y_{i+1}^k, v_{i+1}^k\right)}$$

It should be said that of the four values obtained, $\frac{\partial x_P}{\partial\psi}, \frac{\partial y_i^{k+1}}{\partial\psi}, \frac{\partial v_i^{k+1}}{\partial\psi}$ and $\frac{\partial x_Q}{\partial\psi}$, we are only interested in keeping $\frac{\partial y_i^{k+1}}{\partial\psi}$ and $\frac{\partial v_i^{k+1}}{\partial\psi}$ to find the solution to k+1, as the two other values never intervene in the right-hand side of system [5.33].

Regarding the differences between equations [5.33] and [5.30], we should underline the presence of the matrix $[S]_i^k$, whose values can be obtained from equation [5.32]. We also note that from the known values $\frac{\partial y_{i-1}^k}{\partial \psi}, \frac{\partial v_{i-1}^k}{\partial \psi}, \frac{\partial y_i^k}{\partial \psi}, \frac{\partial v_i^k}{\partial \psi}, \frac{\partial y_{i+1}^k}{\partial \psi}$ and $\frac{\partial v_{i+1}^k}{\partial \psi}$ at time k, the values of $\frac{\partial y_i^{k+1}}{\partial \psi}$ and $\frac{\partial v_i^{k+1}}{\partial \psi}$ at time k+1 are obtained, which shows the concept of influence and dependence domain.

For each point of the structured grid, a set of equations of the type [5.33] can be solved, except for nodes that correspond to checkpoints or target points with pumps, gates and boundary conditions. These cases are analyzed in the following two sections.

5.5.4. *Control structures*

There are many flow control structures in canals that allow flow modeling according to the specification of the watermaster. Individual study of each of these structures is impossible in this work; for this reason, we will introduce the most common structures. A commonly found structure is a checkpoint with a sluice gate, a lateral weir outlet, an orifice offtake and a pump, as seen in Figure 5.16. The interaction of this control structure with the flow can be described according to the principles of mass and energy conservation. These principles establish two mathematical relations between the flow conditions upstream and downstream of the checkpoint (see Figure 5.16):

$$\left.\begin{aligned} S(y_e)\frac{dy_e}{dt} &= A(y_e)v_e - q_b - q_s(y_e) - A(y_s)v_s - q_{offtake}(y_e) \\ A(y_s)v_s &= k_c u\sqrt{y_e - y_s + d} \end{aligned}\right\} \qquad [5.34]$$

where:

– $S(y_e)$ is the horizontal surface of the reception area in the checkpoint.

– $A(y_e)v_e$ is the incoming flow to the checkpoint, defined in terms of water level and velocity.

– $A(y_s)v_s$ is the outgoing flow to the checkpoint, which continues along the canal, described in terms of water level and velocity.

– $k_c=\sqrt{2g}*C_c*a_c$, where C_c is the discharge coefficient of the sluice gate and a_c is the width of the sluice gate.

– d is the checkpoint drop and u is the gate opening.

– q_b is the pumping offtake.

– $q_s(y_e) = C_s * a_s * (y_e - y_0)^{3/2}$ is the outgoing lateral flow through the weir, where C_s is the discharge coefficient, a_s is the width of the lip and y_0 is the height of the lip measured from the bottom.

– $q_{offtake}(y_e) = C_d * A_0 * \sqrt{2g * y_e}$ is the outgoing orifice offtake flow, where C_d is the discharge coefficient and A_0 is the area of the orifice offtake.

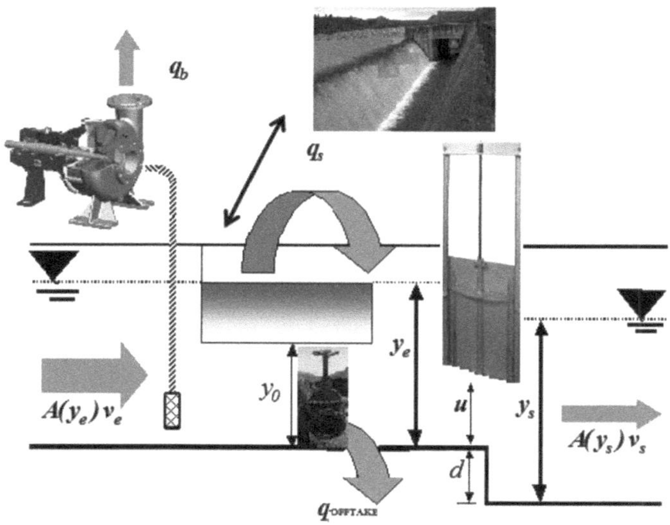

Figure 5.16. *Diagram of a checkpoint with gate, lateral weir and pump*

In canal control, there are several different ways to deliver water from the canal. For instance, the canal operator could pump the water of the canal, so that the control relies exclusively on the operator. Instead, the lateral weir outlet or orifice offtake are not dependent on anyone; these structures get water from the canal for themselves, that is, no operator directly controls the delivered water from the canal, but the water level difference among offtake or lateral weir level and water level is mainly determinant for water delivery (as well as the weir coefficient, etc.). Therefore, in this second case, the control of the extracted flow will be much more difficult. Apart from the cases in which pumping is compulsory (due to level limitations), the overflow system is preferred because it has a lower energy cost even though it is more difficult to control, although this latter case depends on the irrigation canal and the importance of the demanded deliveries.

5.5.5. *Discretization of the checkpoint equations*

The presence of checkpoints or control structures in the middle of a canal leads to the subdivision of this canal into canal pools in such a way that there is a canal pool between two checkpoints and there is a checkpoint between two pools. Therefore, y_n^{k+1} represents the water level at node n in the section upstream of the control point at time k+1, that is, the incoming water level y_e. In the same way, y_1^{k+1} is defined as the existing water level at the first node of the downstream pool from the checkpoint at the same time k+1 and y_s is defined as the outgoing water level at the control point (Figure 5.17). The same can be said for the velocities v_n^{k+1} and v_1^{k+1}.

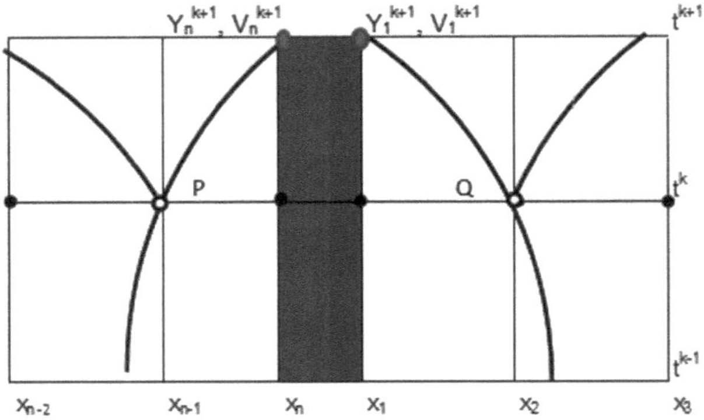

Figure 5.17. *Graph with discretization of the control point equations*

If discretization is carried out with time and we rewrite the control point equations [5.34], join them with the characteristics of equations [5.27] and then change the nomenclature, then the following system of six equations is obtained:

$$\left.\begin{aligned}
&f_1 \equiv x_n - x_P - \frac{1}{2}\Delta t\left[v_n^{k+1} + c_n^{k+1} + v_P + c_P\right] = 0\\
&f_2 \equiv \left(v_n^{k+1} - v_P\right) + \frac{g}{2}\frac{c_n^{k+1} + c_P}{c_n^{k+1} c_P}\left(y_n^{k+1} - y_P\right) - g\Delta t\left(\frac{S_{fn}^{k+1} + S_{f_P}}{2} - S_0\right) = 0\\
&f_3 \equiv \left(v_1^{k+1} - v_Q\right) - \frac{g}{2}\frac{c_1^{k+1} + c_Q}{c_1^{k+1} c_Q}\left(y_1^{k+1} - y_Q\right) - g\Delta t\left(\frac{S_{f1}^{k+1} + S_{f_Q}}{2} - S_0\right) = 0\\
&f_4 \equiv x_1^{k+1} - x_Q - \frac{1}{2}\Delta t\left[v_1^{k+1} - c_1^{k+1} + v_Q - c_Q\right] = 0\\
&f_5 \equiv A\left(y_n^{k+1}\right)v_n^{k+1} - q_b - q_s\left(y_n^{k+1}\right) - A\left(y_1^{k+1}\right)v_1^{k+1} - q_{offtake}\left(y_n^{k+1}\right) = 0\\
&f_6 \equiv A\left(y_1^{k+1}\right)v_1^{k+1} - k_c u\sqrt{y_n^{k+1} - y_1^{k+1}} + d = 0
\end{aligned}\right\} \qquad [5.35]$$

where:

– $\Delta t = t^{k+1} - t_P = t^{k+1} - t_Q$

– $y_P(x_P) = s(x_P, y_{n-2}^k, y_{n-1}^k, y_n^k)$

– $y_Q(x_Q) = s(x_Q, y_1^k, y_2^k, y_3^k)$

– $v_P(x_P) = s(x_P, v_{n-2}^k, v_{n-1}^k, v_n^k)$

– $v_Q(x_Q) = s(x_Q, v_1^k, v_2^k, v_3^k)$

– $c_n^{k+1} = c(y_n^{k+1})$

– $c_1^{k+1} = c(y_1^{k+1})$

– ${S_f}_n^{k+1} = S_f(y_n^{k+1}, v_n^{k+1})$

– ${S_f}_1^{k+1} = S_f(y_1^{k+1}, v_1^{k+1})$

where the unknowns are x_P, y_n^{k+1}, v_n^{k+1}, y_1^{k+1}, v_1^{k+1} and x_Q.

In order to continue with the calculation of the influences of a general parameter ψ, it is necessary to assume, in our case, that this parameter defines the pumping flow $q_b(\psi)$, and that the unknown variables x_P, y_n^{k+1}, v_n^{k+1}, y_1^{k+1}, v_1^{k+1} and x_Q depend on ψ. Therefore, applying again the implicit function theorem to equations [5.35], we obtain:

$$\mathbf{M}\frac{\partial}{\partial\psi}\begin{pmatrix} x_P \\ y_1^{k+1} \\ v_1^{k+1} \\ y_n^{k+1} \\ v_n^{k+1} \\ x_Q \end{pmatrix} = \mathbf{NS}\frac{\partial}{\partial\psi}\begin{pmatrix} y_{n-2}^k \\ v_{n-2}^k \\ y_{n-1}^k \\ v_{n-1}^k \\ y_n^k \\ v_n^k \\ y_1^k \\ v_1^k \\ y_2^k \\ v_2^k \\ y_3^k \\ v_3^k \end{pmatrix} + \mathbf{L}\frac{\partial q_b}{\partial\psi} \qquad [5.36]$$

where:

$$\mathbf{M} = \frac{\partial(f_1, f_2, f_3, f_4, f_5, f_6)}{\partial\left(x_P, y_n^{k+1}, v_n^{k+1}, y_1^{k+1}, v_1^{k+1}, x_Q\right)}$$

$$\mathbf{N} = -\frac{\partial(f_1, f_2, f_3, f_4, f_5, f_6)}{\partial\left(x_P, y_P, v_P, y_Q, v_Q, x_Q\right)}$$

$$\mathbf{L} = \begin{pmatrix} 0 & 0 & 0 & 0 & \frac{\partial f_5}{\partial q_b} & 0 \end{pmatrix}^T$$

$$\mathbf{S} = \frac{\partial\left(x_P, y_P, v_P, y_Q, v_Q, x_Q\right)}{\partial\left(x_P, y_{n-2}^k, v_{n-2}^k, y_{n-1}^k, v_{n-1}^k, y_n^k, v_n^k, y_1^k, v_1^k, y_2^k, v_2^k, y_3^k, v_3^k, x_Q\right)}$$

So far, we have dealt with the influence of a generic variable ψ on the solution without going into detail about what the variable could represent. In equation [5.36], for the first time the pump flow $q_b(\psi)$ explicitly appears in the description. Despite the fact that the specific form of this function is still unknown, equation [5.36] shows that the influence of the parameter ψ on flow conditions at time k+1 is the sum of the indirect influence of the conditions at instant k and the direct influence at instant k+1 through the term $\left|\frac{\partial q_b}{\partial \psi}\right|_n^{k+1}$, which represents the variation in the pump flow when the parameter ψ changes.

5.5.6. *Pump flow trajectories*

In irrigation canal control, the time required by the pump to reach full-power operation can be considered negligible compared to the rest of the time during which the pump remains working.

This is the reason for representing the pump flow trajectory mathematically with a piece-wise function, as shown in Figure 5.6. This representation allows the parameterization of the problem, which is the process of identifying the unknowns. By simply identifying the parameter ψ as the pump flow value $q_b(K)$ at the interval K between the time instants t^{K-1} and t^K, the process is finished. A pump flow trajectory is defined as the following vector of variables:

$$Q_b = \left[q_b(1) \quad \cdots \quad q_b(K) \quad \cdots \quad q_b(K_F)\right]^T \qquad [5.37]$$

where K_F is the last interval in which the function of the pump flow trajectory is defined.

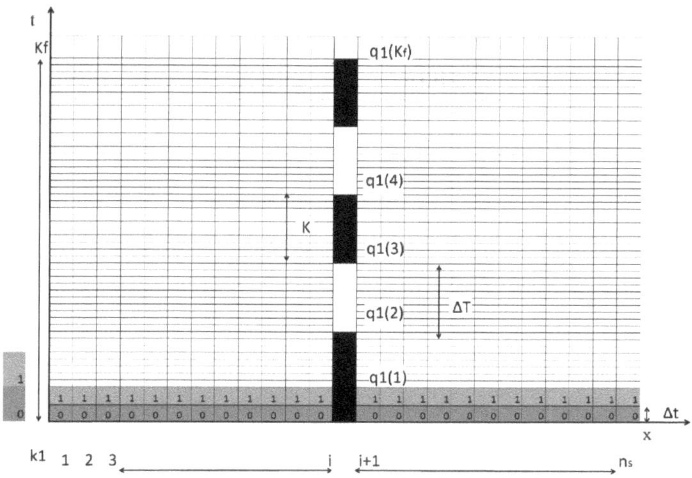

Figure 5.18. *Graphical representation of the time discretization on two canal pools*

A pump flow change implies the following important consequences:

– In order to solve the problem, it will be necessary to use a set of Saint-Venant equations, for example those described in equations [5.27]. It will be also necessary to limit the maximum time step Δt for stability reasons, as described in the stability condition of Courant–Friedrichs–Lewy (CFL). According to this condition, the maximum step allowed is the time taken by a wave to travel the incremental space Δx. If pump parameters were defined for each discretization time instant, then the total number of variables would depend on CFL, that is, on the existing flow conditions for each instant. If it is very difficult, or even impossible, to solve the problem, then it is necessary to establish a time discretization of the pump flow trajectory independent of that of the simulation. To show this difference, the superscript K (uppercase) will be used for the time discretization of the pump flow trajectory and k (lowercase) will be used for the discretization of the simulation, called time step, as shown in Figure 5.18. The time discretization K is usually called the regulation period.

– The piece-wise form of function in Figure 5.6 implies $\frac{\partial q_b}{\partial \psi} = 1$ in equation [5.36], and so we will talk about the pumping flow $q_b(K)$ during the K-interval of time (Figure 5.18).

– Taking into account the pump flow trajectory vector [5.37], the variables are defined in a time window, which allows us to state: $\frac{\partial y_i^k}{\partial q_b(K)} = 0$ if $t^k < T^K$.

5.5.7. *Boundary conditions*

To conclude the study of the influence of pump flow on the flow conditions at a point in the canal at a specific time instant, it is necessary to describe the boundary conditions as well as to determine how the influences evolve when they arrive at the edge of the canal and bounce. Crandall [CRA 56] demonstrates another uniqueness theorem for second-order systems like the one studied in this book, using the graph in Figure 5.19. The theorem states that if the flow conditions at a characteristic intersection S are known and if only one variable is known from the two non-characteristic curves SP and SQ, then the solution uniqueness is guaranteed in the shaded area SPRQ.

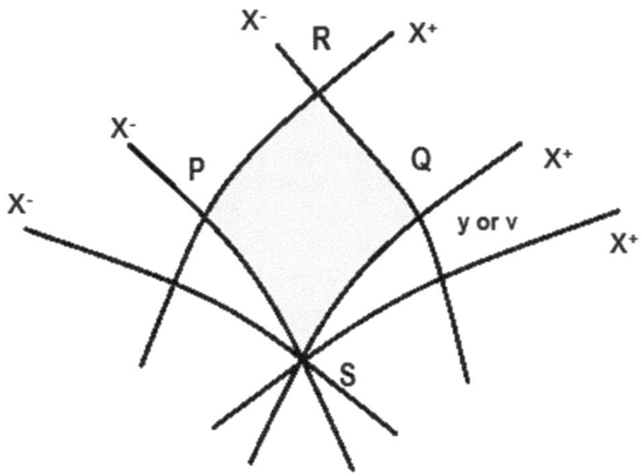

Figure 5.19. *Formulation of the second theory of uniqueness for second-order sets of hyperbolic equations. For a color version of the figure, please see www.iste.co.uk/bonet/canals.zip*

The joint application of this theory and the one mentioned previously (Figure 5.11) entitles us to state that by establishing two conditions, one at each extreme of the solution interval, uniqueness can be guaranteed in the shaded areas of Figure 5.20. That is, if flow conditions at points P and Q on two characteristic curves and any condition of the y axis at both extremes are known, then the solution for points R and R' can be obtained.

Figure 5.20. *Application of the two uniqueness theorems to the boundaries. For a color version of the figure, please see www.iste.co.uk/bonet/canals.zip*

It should be noted that in a subcritical regime (the usual working regime in irrigation canals and the one considered in this book), a condition for each canal boundary must be established (as stated previously) because the slopes of the characteristic curves x+ and x− have opposite signs as the velocity of the wave is higher than the velocity of fluid media. That is, from equations [5.22], we have:

$$\frac{dx^+}{dt} \equiv v + c(y) > 0; \qquad \frac{dx^-}{dt} \equiv v - c(y) < 0$$

A large number of boundary conditions can be set at the extreme ends of a canal. Continuing with the nomenclature of the structured grid (Figure 5.18), the boundary conditions are established as two expressions of the following generic types:

$$\begin{aligned} f_7 &\equiv \left(y_1^{k+1} \quad v_1^{k+1}\right) = 0 \text{ upstream boundary condition} \\ f_8 &\equiv \left(y_{n_S}^{k+1} \quad v_{n_S}^{k+1}\right) = 0 \text{ downstream boundary condition} \end{aligned} \qquad [5.38]$$

Taking the third and fourth equations of set [5.35] and the first of set [5.38] gives a new system that needs to be solved in order to find the following flow conditions at the upstream boundary:

$$\left.\begin{aligned} & f_7 \equiv \left(y_1^{k+1} \quad v_1^{k+1}\right) = 0 \\ & f_3 \equiv \left(v_1^{k+1} - v_Q\right) - \frac{g}{2}\frac{c_1^{k+1} + c_Q}{c_1^{k+1} c_Q}\left(y_1^{k+1} - y_Q\right) - g\Delta t\left(\frac{S_{f1}^{\,k+1} + S_{f_Q}}{2} - S_0\right) = 0 \\ & f_4 \equiv x_1^{k+1} - x_Q - \frac{1}{2}\Delta t\left[v_1^{k+1} - c_1^{k+1} + v_Q - c_Q\right] = 0 \end{aligned}\right\} \qquad [5.39]$$

By doing so, we can find the set of equations corresponding to the downstream ends if we take the first and second equations of set [5.35] and the second of set [5.38], which gives:

$$\left.\begin{aligned}
&f_1 \equiv x_{n_s} - x_P - \frac{1}{2}\Delta t\left[v_{n_s}^{k+1} + c_{n_s}^{k+1} + v_P + c_P\right] = 0\\
&f_2 \equiv \left(v_{n_s}^{k+1} - v_P\right) + \frac{g}{2}\frac{c_{n_s}^{k+1} + c_P}{c_{n_s}^{k+1} c_P}\left(y_{n_s}^{k+1} - y_P\right) - g\Delta t\left(\frac{S_{f n_s}^{\;k+1} + S_{f_P}}{2} - S_0\right) = 0\\
&f_8 \equiv \left(y_{n_s}^{k+1} \quad v_{n_s}^{k+1}\right) = 0
\end{aligned}\right\} \qquad [5.40]$$

Applying once more the hypothesis that all variables are dependent on a pump flow $q_b(K)$ as we did in order to find equations [5.33] and [5.36] for the upstream point, we can write:

$$\mathbf{M}\frac{\partial}{\partial q_b(K)}\begin{pmatrix} x_Q \\ y_1^{k+1} \\ v_1^{k+1} \end{pmatrix} = \mathbf{NS}\frac{\partial}{\partial q_b(K)}\begin{pmatrix} y_1^k \\ v_1^k \\ y_2^k \\ v_2^k \\ y_3^k \\ v_3^k \end{pmatrix} + \mathbf{L} \qquad [5.41]$$

where:

$$\mathbf{M} = \frac{\partial(f_7, f_3, f_4)}{\partial\left(y_1^{k+1}, v_1^{k+1}, x_Q\right)}$$

$$\mathbf{N} = -\frac{\partial(f_7, f_3, f_4)}{\partial\left(y_Q, v_Q, x_Q\right)}$$

$$\mathbf{L} = \left(\frac{\partial f_7}{\partial q_b(K)} \quad 0 \quad 0\right)^T$$

$$\mathbf{S} = \frac{\partial\left(y_Q, v_Q, x_Q\right)}{\partial\left(y_1^k, v_1^k, y_2^k, v_2^k, y_3^k, v_3^k\right)}$$

and for the downstream end:

$$\mathbf{M}\frac{\partial}{\partial q_b(K)}\begin{pmatrix} y_1^{k+1} \\ v_1^{k+1} \\ x_P \end{pmatrix} = \mathbf{NS}\frac{\partial}{\partial q_b(K)}\begin{pmatrix} y_{n_s-2}^k \\ v_{n_s-2}^k \\ y_{n_s-1}^k \\ v_{n_s-1}^k \\ y_{n_s}^k \\ v_{n_s}^k \end{pmatrix} + \mathbf{L} \qquad [5.42]$$

where:

$$\mathbf{M} = \frac{\partial(f_1, f_2, f_8)}{\partial\left(y_{n_s}^{k+1}, v_{n_s}^{k+1}, x_P\right)}; \quad \mathbf{N} = -\frac{\partial(f_1, f_2, f_8)}{\partial(y_P, v_P, x_P)}$$

$$\mathbf{L} = \begin{pmatrix} 0 & 0 & \dfrac{\partial f_8}{\partial q_b(K)} \end{pmatrix}^T; \quad \mathbf{S} = \frac{\partial(y_P, v_P, x_P)}{\partial\left(y_{n_s-2}^k, v_{n_s-2}^k, y_{n_s-1}^k, v_{n_s-1}^k, y_{n_s}^k, v_{n_s}^k\right)}$$

5.5.8. *Hydraulic influence of a pump flow trajectory parameter on the state vector*

The structured grid (Figure 5.18) shows a time–space discretization of a canal with two pools, Pool I subdivided into i−1 cells and Pool II subdivided into n_s-1 cells. The grid also shows the flow solution at time k and at the subsequent time k+1 (every Δt) for all of the n_S sections. The algorithm allows us to sequentially find solutions for all of the computational nodes at k+1, provided the type of each of these nodes is identified, and the algorithm can solve the corresponding set of equations. Table 5.1 summarizes the whole procedure.

	Pool	Node	Simulation
Upstream boundary	I	1	[5.41]
Computational nodes	I	2, . . . ,i−1	[5.33]
Checkpoints	I and II	i,i+1	[5.36]
Computational nodes	II	i+2, . . . ,n_s-1	[5.33]
Downstream boundary	II	n_s	[5.42]

Table 5.1. *Summary of the whole procedure*

So far, we have only discussed the solution of the equations for simulation purposes. However, our final goal is the study of the time–space evolution of the influence of any variable in the trajectory [5.37]. For each time instant, the new value of the influence of any pump flow $q_b(K)$ is calculated at each computational node using some of the aforementioned systems of equations: [5.33], [5.36], [5.41] or [5.42].

These are adequately compiled into a single set that can be represented in the following manner:

$$\frac{\partial \mathbf{x}^{k+1}}{\partial q_b(K)} = \mathbf{A}\frac{\partial \mathbf{x}^{k}}{\partial q_b(K)} + \mathbf{b}\left[\mathbf{x}^{k+1}, q_b(K)\right] \quad [5.43]$$

where $A = A\left[x^{k+1}, x^k, q_b(K)\right]$ is a $(2n_S)\times(2n_S)$ square matrix, with n_S the total number of cross-sections of the simulation according to the nomenclature shown in Figure 5.18, $b[x^{k+1}, q_b(K)]$ is the direct influence vector of $2n_S$ components and x is the state vector of dimension $2n_S$, whose values at instants k and k+1 are:

$$\mathbf{x}^k = \begin{pmatrix} y_1^k & v_1^k & \cdots & \cdots & y_i^k & v_i^k & \cdots & \cdots & y_{n_S}^k & y_{n_S}^k \end{pmatrix}^T \quad [5.44]$$

$$\mathbf{x}^{k+1} = \begin{pmatrix} y_1^{k+1} & v_1^{k+1} & \cdots & \cdots & y_i^{k+1} & v_i^{k+1} & \cdots & \cdots & y_{n_S}^{k+1} & y_{n_S}^{k+1} \end{pmatrix}^T$$

The set of equations [5.43] shows that the influence of the pump parameter $q_b(K)$ on the state vector at time k+1 is equal to the direct influence of $q_b(K)$ at instant k+1 plus the updating of the influence at time k.

Figure 5.21 shows the influence of a pump flow change during several regulation periods (T^K–T^{K+1}), as the disturbance extends the upstream and downstream in the canal. The gray dots represent points where the disturbance has no influence, and the blue dots represent points where the disturbance decreases or increases water levels due to the pump flow. The plot can be also seen as the effect of a perturbation made by $q_b(K)$ over the predictive vector. From the diagram shown in Figure 5.21, we make the following considerations:

a) If we study the influence of $q_b(K)$ at time step k, as $t^k < T^K$, then $\frac{\partial x_i^K}{\partial q_b(K)} = 0$ and there is no need to set the influence of the pump flow because the influence domain of a parameter backward in time is null.

b) If we study the influence of $q_b(K)$ at time step k+1, as $t^{k+1} = T^K$, then the set of equations [5.43] becomes:

$$\frac{\partial x^{k+1}}{\partial q_b(K)} = [A]\frac{\partial x^k}{\partial q_b(K)} + b\left[x^{k+1}, q_b(K)\right] = b\left[x^{k+1}, q_b(K)\right] \qquad [5.45]$$

where $[A] = A[x^{k+1}, x^k, q_b(K)]$ represents the updating of the influence from x^k to x^{k+1}, $b[x^{k+1}, q_b(K)]$ is the direct influence vector and x is the state vector.

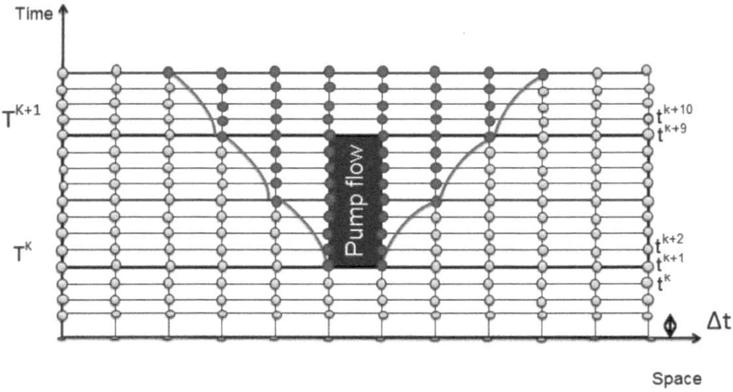

Figure 5.21. *Evolution of the influence of the pump flow through time and space. The gray dots represent the points where there is no influence of the pump flow:* $\frac{\partial X_i^K}{\partial q_b(K)} = 0$*, and the blue dots represent the points where there is influence of the pump flow:* $\frac{\partial X_i^K}{\partial q_b(K)} \neq 0$*. For a color version of the figure, please see www.iste.co.uk/bonet/canals.zip*

This last term is useful in finding the initial value of the influence of the parameter $q_b(K)$ on the flow conditions x^{k+1}, that is, the direct influence.

c) If we study the influence of $q_b(K)$ at time step k+2, then the complete set of equations [5.43] is used when $T^K < t^{k+2} \leq T^{K+1}$:

$$\frac{\partial x^{k+2}}{\partial q(K)} = A\left[x^{k+1}, q_b(K)\right]\frac{\partial x^{k+1}}{\partial q(K)} + b\left[x^{k+2}, q_b(K)\right] \qquad [5.46]$$

d) Finally, when $t^{k+10} > T^{K+1}$, equation [5.43] loses the direct influence of the parameter and becomes:

$$\frac{\partial x^{k+10}}{\partial q(K)} = A\left[x^{k+9}, q_b(K)\right]\frac{\partial x^{k+9}}{\partial q(K)} \qquad [5.47]$$

After determining the values of the state vector by applying the corresponding sets of equations, the influence domain of the parameter $q_b(K)$ can be described and quantified, and its influence on the system is shown by the blue dots in Figure 5.21.

The hydraulic influence matrix is composed of the influence of the pump flow over the canal state gathered in a matrix, which is composed of the following terms:

$$\left[\text{HIM}\left(Q_{b_j}\right)\right] = \left[\frac{\partial x_{k_1}^{k_F}}{\partial q_{b_j}(1)} \cdots \frac{\partial x_{k_1}^{k_F}}{\partial q_{b_j}(K)} \cdots \frac{\partial x_{k_1}^{k_F}}{\partial q_{b_j}(K_F)}\right]. \qquad [5.48]$$

A forecast of the perturbed flow can be calculated by the product:

$$\Delta X_1^{k_F} = \left(\frac{\partial x_1^{k_F}}{\partial q_b(K)}\right)\Delta q_b(K) \qquad [5.49]$$

If we want to know a forecast of the predictive vector when the pump flow trajectories are perturbed a little, we only have to consider:

$$X_1^{k_F}(Q_b{}^* + \Delta Q_b) = X_1^{k_F}(Q^*{}_b) + \Delta X_1^{k_F} \qquad [5.50]$$

where Q_b is the vector of the compilation of all the pump flow trajectories:

$$Q_b = \left(q_{b_1}(1) \dots\dots q_{b_i}(1) \cdots\cdots q_{b_{n_p}}(1) \dots q_{b_1}(K_F) \dots\dots q_{b_i}(K_F) \cdots\cdots q_{b_{n_p}}(K_F)\right) \quad [5.51]$$

where $\Delta Q_b = (0 \dots \Delta q_b(K) \dots)$ is the disturbance of the pump flow trajectories, Q^*_b is the reference pump flow trajectories and $X_1^{k_F}(Q_b{}^* + \Delta Q_b)$ is the simulated state vector when the pump flow trajectories are disturbed a little.

The HIM(Q_b) allows the analysis of the effect of a flow change on the system (which we could call analysis of the "influence domain") and also which pump station has an effect on a cross-section of the canal at a certain instant (which we could call analysis of the "dependence domain" of the point).

The HIM(Q_b) can be considered as a linearization of the Saint-Venant model as the model used by Malaterre [MAL 94], but the HIM(Q_b) is a linearization model of an unsteady-state solution. The HIM is a helpful tool to calculate the state vector when the pump flow trajectories are disturbed a little:

$$X_1^{k_F}(Q_q{}^* + \Delta Q_q) - X_1^{k_F}(Q_q{}^*) = \mathrm{HIM}(Q_q{}^*)\Delta Q_q + O(\Delta Q_q{}^2) \qquad [5.52]$$

Neglecting the second-order and upper terms, equation [5.52] can be rewritten in the form:

$$\Delta X = \mathrm{HIM}(Q_q{}^*)\Delta Q_q \qquad [5.53]$$

where ΔQ_b is the perturbed pump flow vector and ΔX is the predicted error from the state vector.

We may include all of the state vectors [5.44] for each k-instant on the simulated past time horizon in the following single vector, called the predicted vector:

$$\mathbf{X}_1^{k_F}(Q) = \left[\mathbf{x}^1(Q)^T \quad \mathbf{x}^2(Q)^T \quad \ldots \quad \mathbf{x}^{k_F-1}(Q)^T \quad \mathbf{x}^{k_F}(Q)^T\right]^T \qquad [5.54]$$

where k_F is the final instant of the past time horizon of a period of time. We can now compile into only one vector, in a similar way to before, all of the values of the influence of any pump flow trajectory $q_{bj}(K)$ on the state vector. We will call this new vector the vector of influence of the pump flow on the predictive vector:

$$\frac{\partial \mathbf{X}_1^{k_F}(Q)}{\partial q_{bj}(K)} = \left[\frac{\partial \mathbf{x}^1(Q)^T}{\partial q_{bj}(K)} \quad \frac{\partial \mathbf{x}^2(Q)^T}{\partial q_{bj}(K)} \quad \ldots \quad \frac{\partial \mathbf{x}^{k_F-1}(Q)^T}{\partial q_{bj}(K)} \quad \frac{\partial \mathbf{x}^{k_F}(Q)^T}{\partial q_{bj}(K)}\right]^T \qquad [5.55]$$

5.5.9. *Compilation of the hydraulic influence matrix*

5.5.9.1. *Definition*

After introducing the hydraulic influence of a pump flow on the canal state, it is possible to compile all these elements into a matrix, which we call "the hydraulic influence matrix". If we define the hydraulic influence on state vector [5.54] of a certain pump station “j” ([5.37]) during a past time horizon (γ) in each cross-section, then the following hydraulic influence matrix, denoted by $[I_M(q_b)]_j$, is obtained:

$$[\mathrm{I_M}(q_b)]_j = \left[\frac{\partial X_{k_1}^{k_F}}{\partial q_{b_j}(1)} \quad \cdots \quad \frac{\partial X_{k_1}^{k_F}}{\partial q_{b_j}(K)} \quad \cdots \quad \frac{\partial X_{k_1}^{k_F}}{\partial q_{b_j}(\gamma)}\right] =$$

$$\begin{bmatrix} \dfrac{\partial x^{k_I}}{\partial q_{b_j}(1)} & \cdots & \dfrac{\partial x^{k_I}}{\partial q_{b_j}(K)} & \cdots & \dfrac{\partial x^{k_I}}{\partial q_{b_j}(\gamma)} \\ \vdots & \ddots & \vdots & \ddots & \vdots \\ \dfrac{\partial x^{k_i}}{\partial q_{b_j}(1)} & \ddots & \dfrac{\partial x^{k_i}}{\partial q_{b_j}(K)} & \ddots & \dfrac{\partial x^{k_i}}{\partial q_{b_j}(\gamma)} \\ \vdots & \ddots & \vdots & \ddots & \vdots \\ \dfrac{\partial x^{k_F}}{\partial q_{b_j}(1)} & \cdots & \dfrac{\partial x^{k_F}}{\partial q_{b_j}(K)} & \cdots & \dfrac{\partial x^{k_F}}{\partial q_{b_j}(\gamma)} \end{bmatrix} \qquad [5.56]$$

where $2\times n_s$ is the dimension of the state vector at time step k equal to the number of cross-sections plus the number of parameters (water level and velocity), j is the “j^{th}” pump station in the canal and γ is the final regulation period.

A feature of this matrix is the number of columns and rows. There are more rows than columns because every term $\frac{\partial x^k}{\partial q_{b_j}(K)}$ has $2\times n_s$ rows, and the temporal discretization of the simulation Δt^k is smaller than the regulation period ΔT^K.

If the matrix only defines the hydraulic influence of the pump flow on the water level (not considering the velocity), then the number of elements by each term $\frac{\partial x^k}{\partial q_{b_j}(K)}$ is reduced to n_s, and if the number of cross-sections is also reduced to the number of checkpoints, then $n_s = n_c$. If the temporal discretization of the simulation introduced into the matrix is equal to the number of regulation periods and the number of checkpoints is equal to the number of pumps, then the HIM is a square matrix.

Another feature of the HIM is that there are some elements with a zero value, because there is no influence of the pumps on certain points at a certain time step.

When there is more than one pump station in the canal, the matrix has to compile the pump stations all together. In that case, the HIM is compiled as:

$$[\mathrm{I_M}(Q_b)]_X = \left[\left[\frac{\partial X_1^{k_F}(q_b)}{\partial q_{b_1}(1)} \cdots \frac{\partial X_1^{k_F}(q_b)}{\partial q_{b_{np}}(1)}\right] \cdots \left[\frac{\partial X_1^{k_F}(q_b)}{\partial q_{b_1}(\gamma)} \cdots \frac{\partial X_1^{k_F}(q_b)}{\partial q_{b_{np}}(\gamma)}\right]\right] \quad [5.57]$$

where the sub-index X is used to denote the hydraulic influence matrix on the state vector and n_p is the total number of pump stations.

5.5.9.2. *Discrete observer*

In most cases, we only need to know some values of the state vector at specific points of the canal, particularly time steps. It is possible to see an example of two canal pools with one pump in Figure 5.18. In this spatial discretization, the number of sections is n_s and the number of time steps (remember that the duration of the time step depends on the Courant condition in implicit schemes) is $n_T = k_F - k_I$. The discretization of the pump flow trajectories depends on the number of pumps, $n_p = 1$, and the regulation periods, in this case, $\gamma = 4$. As a result, the total number of parameters becomes $n_Q = \gamma \times n_p = 4$.

Therefore, the dimensions of the vectors and matrices shown in this spatial discretization are the following:

– The dimension of the simulation vector $X_{k_{I+1}}^{k_F}$: $n_x = 2 \times n_T \times n_s$;

– The dimension of the pump flow vector Q_b: $n_Q = \gamma \times n_p = 4$;

– The dimension of the hydraulic influence matrix on the simulation vector $[I_M(Q_b)]_x$: $n_x \times n_Q = 2 \times n_T \times n_s \times n_Q$;

– The result of the simulation at checkpoints $Y_{k_{I+1}}^{k_F}(Q_b)$: $n_Y = n_T \times n_s$;

– The dimension of the hydraulic influence matrix on the discrete observer set $[I_M(Q_b)]_Y$: $n_Y \times n_Q$.

In order to express this information mathematically, as we introduced before, a new matrix needs to be introduced, called the “discrete observer matrix” and denoted as [C] in the control literature. This matrix is exclusively made up of “zeros” and "ones", and in equation [5.58] it represents the state vector at certain time steps and at certain points of the canal, as well as the influence matrix of the trajectories at these points.

$$Y_1^{k_F}(Q_b) = [C]X_1^{k_F}(Q_b)$$
$$[I_M(Q_b)]_Y = [C][I_M(Q_b)]_x \qquad [5.58]$$

When $n_T = \gamma, n_s =$ number of checkpoints $= n_c$ and the only hydrodynamic variable represented is the water level, the HIM is a square matrix; this is the conclusion reached previously.

5.5.9.3. *Features of the hydraulic influence matrix*

One of the most important characteristics of the hydraulic influence matrix ($[I_M\ (Q_b)]_Y$) is the fact that it is not a non-singular matrix. This requisite is justified by the fact that in the optimization process, the matrix has to be inverted through its pseudo-inversion, equation [5.59] (see the algorithms from section 5.3.2). In other words, if we want to know what pump flow change (ΔQ_b) has to be introduced into the pump flow trajectory Q_b to obtain a change of the state vector (ΔY), it is necessary to solve the pseudo-inversion, as we introduce in Appendix 2 (section 5.6):

$$\Delta Q_b = \left\{\left[I_M(Q_b)\right]_Y^T \left[I_M(Q_b)\right]_Y\right\}^{-1} \left[I_M(Q_b)\right]_Y^T \Delta Y \qquad [5.59]$$

where $\{[I_M(Q_b)]_Y{}^T [I_M(Q_b)]_Y\}^{-1}$ is the so-called pseudo-inverse matrix and $\Delta Y = Y_1^{k_F}(Q_b + \Delta Q_b) - Y_{k_1}^{k_F}(Q_b)$ is a deviation on the water level. Therefore, if the hydraulic influence matrix is not singular, then the pseudo-inversion will be able to be inverted and the set of equations [5.59] can be solved.

In some cases, the hydraulic influence matrix becomes a singular matrix because a flow extraction from pump station "J" (q_{jb} (K_F –1) at time period K_F –1 has no influence at section "i" during time window (1 to K_F), that is, the wave perturbation at "j" will arrive at section "i" later than time step K_F (see Figure 5.22).

$$\frac{\partial Y_1^{k_F}(Q_b)}{\partial q_{b_j}(K)} = 0 \qquad [5.60]$$

As can be seen in Figure 5.22, the parameter q_{bj} (K_F) has no influence on the levels $Y_i(1),\ldots,Y_i(K_F - 2)\ldots Y_i(K_F - 1)\ldots\ Y_i(K_F)$. As a conclusion, the study of the hydraulic influence matrix could allow the establishment of the control parameters, such as the determination of the testing period.

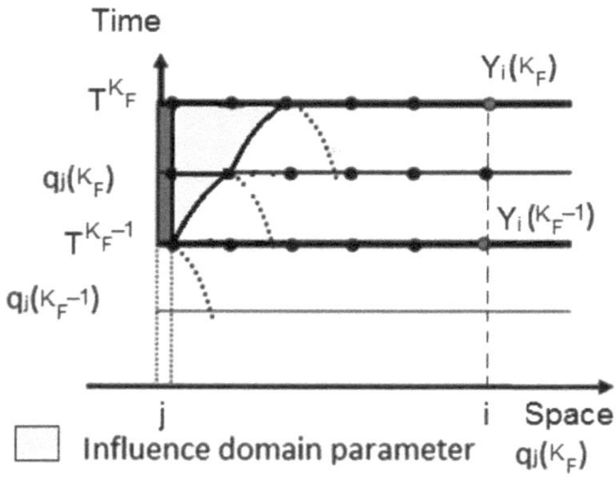

Figure 5.22. *Example of a pump parameter that has no influence on any point. For a color version of the figure, please see www.iste.co.uk/bonet/canals.zip*

5.6. Appendix 2

5.6.1. *Unconstrained optimized problem (Levenberg–Marquardt)*

As shown in the "perturbation theory" by Brogan [BRO 85] applied in the predictive output vector, we can introduce Taylor series, which modifies this vector around a pump flow reference trajectory by means of:

$$Y = Y(Q_b^* + \Delta Q_b) = Y^*(Q_b^*) + \left[\frac{\partial Y}{\partial Q_b}\right]_{Q_b^*} \cdot \Delta Q_b + \frac{1}{2}\left[\frac{\partial^2 Y}{\partial Q_b^{\,2}}\right]_{Q_b^*} \cdot \Delta Q_b^{\,2} + \cdots \quad [5.61]$$

Assuming that the disturbance caused is sufficiently small, we can neglect the terms after the first term in [5.62], obtaining:

$$\Delta Q_b = Y(Q_b^* + \Delta Q_b) - Y^* = \left[\frac{\partial Y}{\partial Q_b}\right]_{Q_b^*} \Delta Q_b \quad [5.62]$$

On the contrary, it is possible to use the hydraulic influence matrix (HIM), defined by Soler [SOL 03]:

$$\left[J^*\right] = \left[\frac{\partial Y}{\partial Q_b}\right]_{Q_b^*} \quad [5.63]$$

Taking into account all of these terms, we can write:

$$\Delta Q_b = \left[J^{*T} \cdot J^{*}\right]^{-1} \cdot \left[J^{*T}\right] \cdot \Delta Y \qquad [5.64]$$

We can set the ΔQ pump flow changes to reach the desired water level in the targets, as these have been modified for a perturbation ΔY. The equation system can be solved by the Jacobi method, the Cholesky method, the singular value decomposition method or LU decomposition. There is extensive literature on these methods, such as *Numerical Recipes in Fortran 77* [PRE 97a].

The problem is to solve equation [5.64] at each time step. We can define the equation in a compact form:

$$[G] = \left[J^{*T}\right] \cdot \left[J^{*}\right] \qquad [5.65]$$

$$b' = \left[J^{*T}\right] \cdot \Delta Y \qquad [5.66]$$

Therefore, we can substitute equations [5.65] and [5.66] into equation [5.64]:

$$b' = [G] \cdot \Delta Q_b \qquad [5.67]$$

Although the system matrix is positive definite by definition, it is not well-conditioned [SOL 03]. This implies that when solving the system, numerical errors can be very important. To avoid this problem, we add a positive term to the diagonal of the system matrix, which is called the Marquardt coefficient (ν). This has been used several times in all problems with the Hessian matrix. Several theories about it exist, for instance in Fletcher [FLE 87].

The system of equations can then be written as:

$$b' = [G + \nu I] \cdot \Delta Q_b \qquad [5.68]$$

where [I] is the identity matrix of the same dimensions as [G]. To solve system [5.68], we can use several different methods. In our case, we use a process optimization method based on the Levenberg–Marquardt method.

We suppose a diagonal matrix [Λ] with the eigenvalues of the matrix [G] and an orthonormal matrix ([V]: [V× V^T] = [I]) defined by the eigenvectors associated with each eigenvalue of [G]. Multiplying both sides of equality [5.68] by the matrix [V^T], we obtain:

$$\left[V^T \cdot G + V^T \cdot \nu I\right]\left[V \cdot V^T\right]\Delta Q_b = \left[V^T\right] \cdot b' \qquad [5.69]$$

Knowing that the matrix [V] is orthonormal, we can write equation [5.69] as follows:

$$\left[V^T \cdot G \cdot V + V^T \cdot \nu I \cdot V\right]\left[V^T\right]\Delta Q_b = \left[V^T\right] \cdot b' \qquad [5.70]$$

And according to the property [Λ] = [V^T×G×V], equation [5.70] can be transformed as follows:

$$[\Lambda + \nu I] \cdot \left[V^T\right] \cdot \Delta Q_b = \left[V^T\right] \cdot b' \qquad [5.71]$$

Furthermore, the fact of using a matrix as orthonormal matrix [V] allows us to ensure that:

$$\left\|[\Lambda + \nu I] \cdot \left[V^T\right] \cdot \Delta Q_b\right\| = \left\|\left[V^T\right] \cdot b'\right\| \qquad [5.72]$$

To simplify equation [5.71], we can define the vectors Z and B as follows:

$$Z = \left[V^T\right] \cdot \Delta Q_b \qquad [5.73]$$

$$B = \left[V^T\right] \cdot b' \qquad [5.74]$$

Therefore, Z is an unknown quantity and equation [5.71] can be transformed into:

$$[\Lambda + \nu I] Z = B \qquad [5.75]$$

From equation [5.75], the Z components can be separated easily because the new system matrix is now diagonal. Each of the Z components can be calculated as follows:

$$Z(i) = \frac{B(i)}{\Lambda(i,i) + \nu[\mathrm{I}]} \qquad [5.76]$$

Although the matrix [G] is positive definite, it is ill-conditioned, and so it is possible that some eigenvalues of [G] are negative, though very close to absolute zero. To prevent this problem, the Marquardt coefficient is introduced, the value of which is necessarily greater than the absolute value of the smallest eigenvalue of [G], so that (Λ+ ν [I]) is always a positive number. Depending on the value chosen for the coefficient, the solution will be different. In this sense, we solve the system of equations through an iteration procedure, and the algorithm reduces the Marquardt coefficient in each iteration in order to reduce the influence of the coefficient in the final solution.

Solving equation [5.76], we obtain the solution (the changes in extracted flow vector):

$$\Delta Q_b = [V] \cdot Z \qquad [5.77]$$

Therefore, an extracted flow vector is calculated at time step K for a past time horizon ($Q_0{}^K{}_b = Q_0{}^K{}_b + \Delta Q_0{}^K{}_b$). The optimization problem must be solved again in the next time instant K + 1, so we could obtain a new extracted flow vector at time step K + 1 for a past time horizon ($Q_0{}^{K+1}{}_b = Q_0{}^{K+1}{}_b + \Delta Q_0{}^{K+1}{}_b$).

6

CSI Performance: Numerical Examples

In this chapter, we test the CSI algorithm on several numerical examples to establish the disturbances introduced into the system and the hydrodynamic state of a canal during a past time horizon.

As demonstrated in Chapter 6, the CSI is a useful algorithm, which can be used in different ways. For instance, if the demand deliveries are increased because a farmer took a greater flow from the orifice offtake than scheduled, then the CSI algorithm can obtain the unknown flow withdrawn by the farmer. Even if a farmer pumps water from an unknown point in the canal, then the algorithm can determine the approximate location of this point and the volume of water withdrawn by the farmer.

The CSI algorithm estimates the hydrodynamic state of a canal during a past time horizon and the current canal state, which is very useful for a feedback controller that needs information about the canal state in real time.

In this chapter, we introduce some numerical examples considering an irrigation canal in subcritical flow regime during a past time horizon, in which the unknown flow changes are associated with changes in demand deliveries at the control structures. The objective is to define the past and current hydrodynamic states of the canal and the unknown flow diversions using the CSI algorithm.

To begin this chapter, we have to define the problem statement, specifying the geometry of the canal, the initial and boundary conditions and the measured water levels at the checkpoints. We analyze the results obtained in each example, that is, the extracted flow water in space and time and the hydrodynamic canal state.

6.1. Introduction

The first objective of the CSI algorithm is to estimate the offtake changes, and the second objective is to obtain the hydrodynamic canal state from the present moment to a past time, for two reasons: to quantify the wrong deliveries and to define the past and current canal states to send to the feedback algorithm. The algorithm calculates the extracted flow (through lateral weirs, pumps, orifice offtake, etc.) from the water-level measurements at checkpoints, the gate trajectories and the scheduled demands at the orifice offtakes during a past time horizon in these examples. This is difficult to achieve because farmers can extract flow from any pump station or orifice offtake in the canal at any time.

We propose several scenarios to test the CSI algorithm in a canal, which has two pools separated by sluice gates (Figure 6.1). The flow is controlled by a gate downstream from the reservoir. Water is delivered through gravity outlets at the downstream end of each pool, where the checkpoints are located. There are pumping stations at the end of each pool, which can introduce disturbances into the system in space and time.

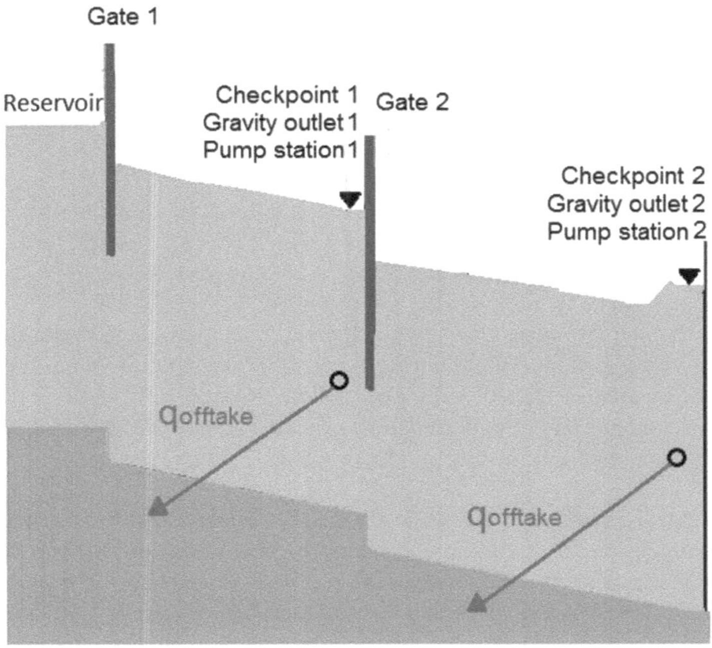

Figure 6.1. *Canal profile*

6.2. Geometry

A canal with a trapezoidal section is shown in Figure 6.1, and the data collected are shown in Table 6.1. Checkpoints, sluice gates, pump stations and orifice offtakes are shown in Table 6.2.

Pool number	Pool length (km)	Bottom slope (%)	Side slopes (H:V)	Manning's coefficient (n)	Bottom width (m)	Canal depth (m)
I	2.5	0.1	1.5:1	0.025	1	2.5
II	2.5	0.1	1.5:1	0.025	1	2.5

Table 6.1. *General canal features*

Number of control structure or checkpoint	Gate discharge coefficient	Gate width (m)	Gate height (m)	Step (m)	Discharge coeff./diameter orifice offtake (m)	Orifice offtake height (m)	Lateral spillway height (m)	Lateral spillway width (m)/discharge coefficient
0*	0.61	5.0	2.5	0.6	–	–	–	–
1	0.61	5.0	2.5	0.6	0.6/0.77	1.0	2.3	500/1.99
2	–	–	–	–	0.6/0.77	1.0	2.3	500/1.99

*There is only a gate in section 0, and there is no checkpoint.

Table 6.2. *Checkpoints and sluice gate/pump station/orifice offtake features (control structure)*

6.3. Discretization of a problem

The past time horizon length is defined as T^{KF} = 14,400 s = 240 min = 4 h. The time interval between successive control actions is ΔT = 300 s = 5 min. The number of regulation periods during the past time horizon is γ = 48. The canal has been spatially discretized (Δx) every 20 m, and therefore, the number of computational cross-sections in each pool is 125 and that along the canal is 250.

6.4. Boundary and initial conditions

In these examples, we consider a large upstream reservoir, whose water level $H_{reservoir}$ is constant (3 m) throughout the test. This is the upstream boundary condition. At the end of the last pool, there is a control structure with orifice offtake and a pump station. The flow through the orifice offtake depends on the upstream water level of the orifice, and the flow pumped is variable in every example. This is the downstream boundary condition. The examples start from an initial steady state with a constant demand delivery at the end of the pools (5 m^3/s through the orifice offtake), and the flow is not pumped at the beginning of the examples. The initial conditions are the same in all examples.

Control structure	Initial flow rate (m^3/s)	Control structure	Initial water level (m)
Gate 1	10.0	Checkpoint 1	2.0
Gate 2	5.0	Checkpoint 2	2.0

Table 6.3. *Initial conditions in the canal*

Control structure	Flow delivered by an orifice offtake (m3/s)
Gravity outlet 1	5.0
Gravity outlet 2	5.0

Table 6.4. *Flow delivered by an orifice offtake at the initial time step*

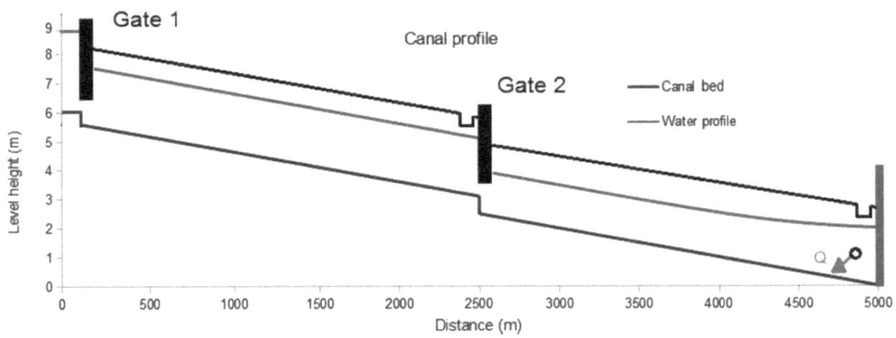

Figure 6.2. *Backwater profile of the canal in the initial steady state. For a color version of the figure, please see www.iste.co.uk/bonet/canals.zip*

The CSI algorithm computes the unscheduled offtake changes introduced into the canal and writes these offtake changes as a pump flow trajectory (temporal function) associated with a certain pump station. The algorithm assigns to the pump stations all flow changes between the initial steady state and the current state.

After introducing the input parameters used in the CSI algorithm as the boundary and initial conditions, we have to define important variables in flow, such as the scheduled deliveries, gate trajectories, pump flow trajectories and measured water levels at the checkpoints during the past time horizon.

– Scheduled deliveries ($q_{offtake}$): there are orifice offtakes to convey the flow rate to the farmers at the end of every pool. The water demand in all scenarios (shown in Table 6.4) was estimated at every regulation time step during the past time horizon and this information is known by the CSI algorithm.

$$Q_{offtake} = Cd \times \frac{\pi \times \emptyset^2}{4} \sqrt{2g(y - y_0)} \qquad [6.1]$$

where y is the water level in the canal at offtake, y_0 is the orifice offtake height equal to $y_{target}/2$, C_d is the discharge coefficient and Ø is the orifice offtake diameter.

– Measured water levels (Y*): the water level is measured at every checkpoint and every time step during the past time horizon. We have to introduce the water-level measurements to CSI for each of these examples.

– Gate trajectories (U): we have to define the sluice gate trajectories during the simulating horizon. The gate position is defined at every regulation period during the past time horizon, and it remains constant in these examples.

– Pump flow trajectories (q_{pump}): CSI has to compute the pump flow trajectories during the past time horizon. The pump flow trajectory vector is composed of two elements (two pump stations) in a simulated time horizon of 48 regulation periods, so the vector size is 96 parameters.

The CSI algorithm solves an optimization problem to set the pump flow trajectories. The algorithm uses an iterative method (Levenberg–Marquardt), which provides successive approximations to the solution, and it is necessary to establish a first pump flow trajectory. In this respect, the Levenberg–Marquardt method is derived from Newtonian methods used in optimization problems and it obtains an accurate solution although the first approximation starts very far from the final minimum. If the first approximation to the solution is close to the optimum solution, then the method converges faster.

6.5. Test cases analyzed

In order to test the CSI algorithm, we introduce different disturbances into the canal in each example, which are unknown to the algorithm. Consequently, we will obtain variations between the measured water level and the expected water level at every checkpoint during the past time horizon. In this sense, if we introduce a pump flow change into a pump station during a period of time, then the initial steady-state conditions of the canal are disturbed and the flow conditions become unsteady. At the end of the test, the canal attains steady state again because the duration of the pump flow disturbance is short.

6.5.1. *First example: flow disturbance*

Water-level measurements (Figure 6.3) are obtained after introducing a flow disturbance of 2 m^3/s (on pump station 1 for 15 min, from 40th min to 55th min). These disturbances are introduced to the computer model as a pump flow change and we obtain the water-level measurements from this model. After the introduction of the water-level measurements into the CSI algorithm, the algorithm proposes the pump flow trajectories that most accurately describe the variation in water levels at the checkpoints during the past time horizon, which lasts 4 h.

The flow disturbance reduces the water level at Checkpoint 1, from 2 m to 1.60 m, and at Checkpoint 2, from 2 m to 1.92 m (see Figure 6.3). A flow change of 2 m^3/s at pump station 1 has a significant impact on the canal profile, and the water level at Checkpoints 1 and 2 returns to the desired level (2 m) in 160 and 150 min, respectively. As soon as both checkpoints recover the desired water level, the flow through the orifice offtake becomes 5 m^3/s again.

The flow disturbance introduced in this test is always an outlet pump flow (positive value), but the algorithm also works well when the pump flow is an inflow or an external runoff (negative value), as we will show in Chapter 10.

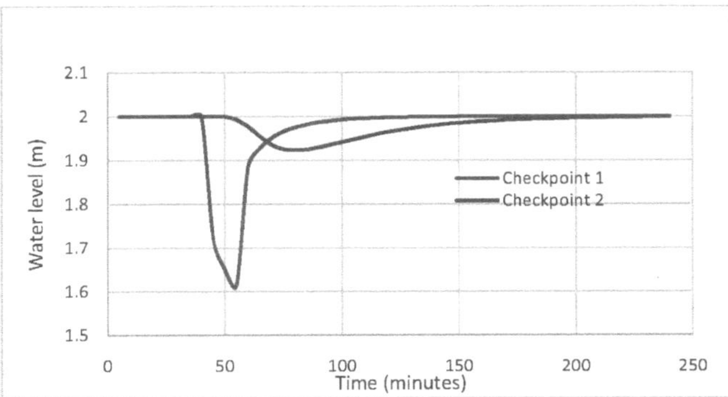

Figure 6.3. *Water level measured at Checkpoints 1 and 2 during the past time horizon in the first example. For a color version of the figure, please see www.iste.co.uk/bonet/canals.zip*

The results of the pump flow obtained by the CSI algorithm are shown in Figure 6.4. We can compare the results between the real pump flow (disturbance) and the simulating pump flow calculated by the algorithm. Both curves match exactly, and differences between them are practically non-existent. The maximum error is about 0.005 m^3/s, and taking into account that the real pump flow is 2 m^3/s, the maximum error percentage between the real and simulated results is 0.25%.

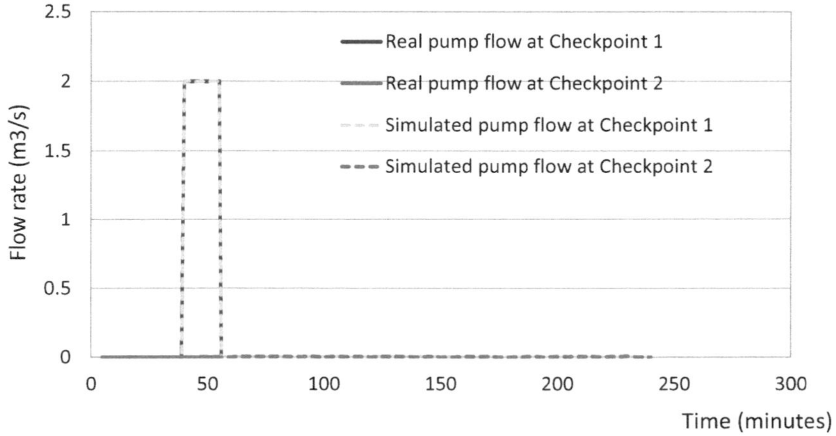

Figure 6.4. *Pump flow at Checkpoints 1 and 2 during the past time horizon in the first example. For a color version of the figure, please see www.iste.co.uk/bonet/canals.zip*

6.5.2. *Second example: small disturbance*

The objective of this test is to set the accuracy of the CSI algorithm in the case of small water offtakes. We analyze the algorithm in a test in which the disturbance is very small. The difference between the real disturbance and the results obtained by the algorithm should be very small, but it is higher in this example because the disturbance is reduced. In cases of small disturbances, we might not be able to differentiate between the real disturbance and the CSI error. If the error committed by the CSI algorithm is much lower than the value of the real disturbance, then CSI is an accurate algorithm.

We introduce a disturbance much lower than the disturbance introduced in the last example and compare the CSI error to the value of the disturbance introduced. In this respect, the disturbance is 50 l/s (1% of the total flow rate in the canal). This disturbance is introduced at pump station 1 for 15 min, from 40th min to 55th min.

In order to illustrate this example, if the error in the result is close to 25 l/s, 0.5% of the total flow rate in the canal, then the error would be similar to the disturbance, and the accuracy of CSI would be low in this test.

The disturbance reduces the water level at Checkpoint 1 from 2 m to 1.990 m, and at Checkpoint 2 from 2 m to 1.998 m (see Figure 6.5). A flow change of 50 l/s at pump station 1 will have a little impact on the water profile of the canal, and the water level at Checkpoints 1 and 2 recovers the desired water level in 50 and 100 min, respectively.

The water-level measurements used by the CSI algorithm are shown in Figure 6.5.

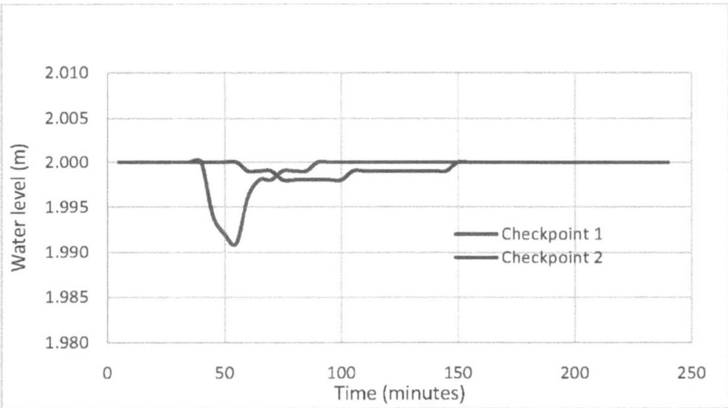

Figure 6.5. *Water level measured at Checkpoints 1 and 2 during a past time horizon in the second example. For a color version of the figure, please see www.iste.co.uk/bonet/canals.zip*

The pump flow results obtained by the CSI algorithm are shown in Figure 6.6. In all cases, the maximum error between the real and simulated solutions is about 0.004 m^3/s; as the real pump flow is 50 l/s, the percentage of maximum error is close to 8%. In this sense, the result obtained by the CSI algorithm is a good approximation to the real pump flow.

In this case, the only problem is having a water level gage with such accuracy, although the gage should be more or less accurate depending on the geometry and features of the canal and the disturbance. In this example, an external disturbance value of 50 l/s represents a water-level disturbance of a few millimeters, and this water-level variation can be too low to be measured accurately by a water gage.

Taking into account the maximum error obtained by the CSI algorithm and the accuracy of the gage, the former is accurate enough for the commonly used gages.

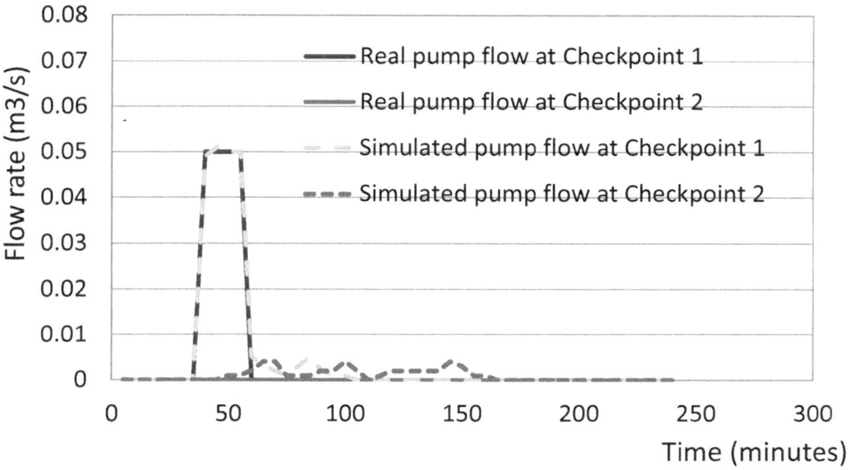

Figure 6.6. *Pump flow at Checkpoints 1 and 2 during a past time horizon in the second example. For a color version of the figure, please see www.iste.co.uk/bonet/canals.zip*

6.5.3. *Third example: flow disturbance not localized at a control structure*

The CSI algorithm calculates pump flow changes at every control structure, but sometimes the pump unit can be located in places other than a known control structure. For this reason, we establish this scenario to check the behavior of CSI when a disturbance is introduced outside of a control structure.

In this example, we introduce a disturbance at a certain cross-section of a canal (P.K. 3 + 750), where the pump flow change is located at the middle of the second pool. This cross-section is *not a control structure*. The CSI algorithm does not have any water level measures at P.K. 3 + 750, as this cross-section is not a node set by the algorithm as a checkpoint and therefore it is a "normal" node for the algorithm.

The water-level measurements (Figure 6.7) that are used as input data in the CSI algorithm are obtained after introducing into the computer model a flow disturbance of 2 m^3/s (at P.K. 3 + 750 for 60 min, from 60th min to 120th min).

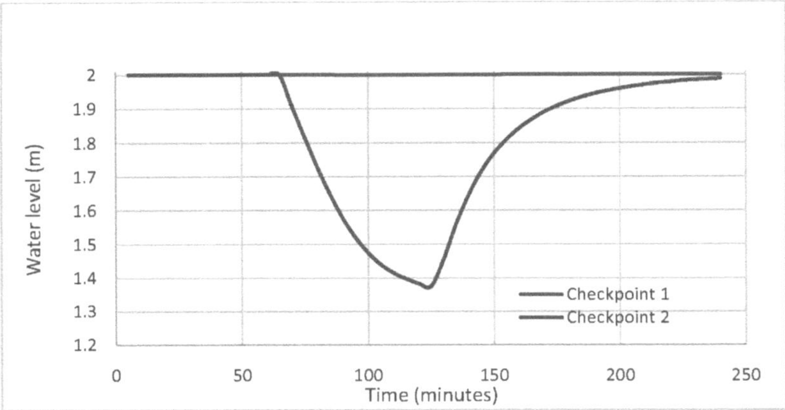

Figure 6.7. *Water level measured at Checkpoints 1 and 2 during a past time horizon in the third example. For a color version of the figure, please see www.iste.co.uk/bonet/canals.zip*

The flow disturbance reduces the water level at Checkpoint 2 from 2 m to 1.375 m, and at Checkpoint 1, the water level is kept constant due to the canal features and the flow conditions (see Figure 6.7). In this sense, a pump flow of 2 m^3/s at P.K. 3 + 750 will have a significant impact on Checkpoint 2 and little impact on Checkpoint 1. The water level at Checkpoint 2 is recovered to the desired water level (2 m) in 130 min. The flow through the second orifice offtake becomes 5 m^3/s at the end of the test.

In this example, the CSI algorithm has to adapt a new situation because the algorithm is only able to calculate disturbances at a control structure, but in this case, the flow rate pumping is not localized at a control structure, but in an unknown cross-section in the canal, between two checkpoints.

The results obtained by the CSI algorithm are shown in Figure 6.8. It distributes the pumped flow in an unknown cross-section to the upstream and downstream pump stations.

The simulated pump flow at Checkpoint 2 (P.K. 5 + 000) starts 5 min after the real pump flow (P.K. 3 + 750) starts pumping. This is because, until the depressive translatory wave reaches Checkpoint 2, 1,250 m downstream from PK 3 + 750, there is no change in the water-level measurements at Checkpoint 2, so the CSI algorithm does not have any notice about the pump flow change. This 5 min difference corresponds to the time that the wave takes to travel from P.K. 3 + 750 to P.K. 5 + 000.

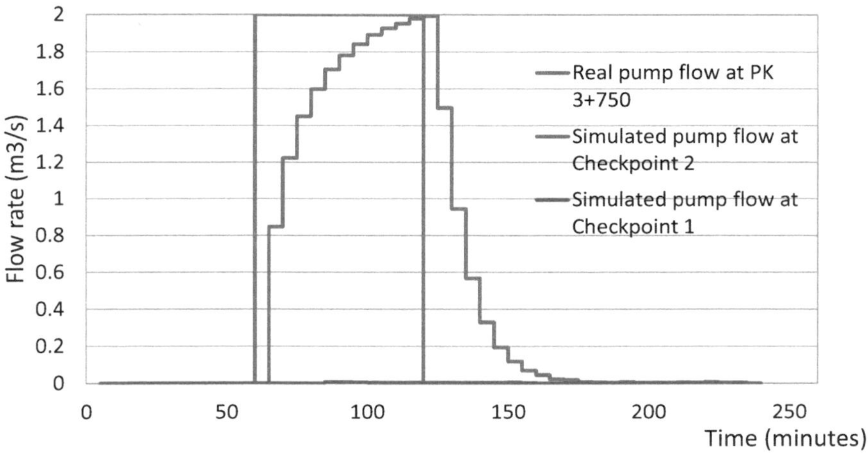

Figure 6.8. *Pump flow at Checkpoints 1 and 2 during a past time horizon in the third example. For a color version of the figure, please see www.iste.co.uk/bonet/canals.zip*

The travel time was also computed by equation [6.2], and the value was approximately 287 s, with an average velocity of 0.54 m/s and an average celerity of 3.80 m/s obtained from the computer model.

$$Time = \frac{(X_{P.K.5+000} - X_{P.K.3+750})}{(Velocity + Celerity)} \approx 287 \ seconds \qquad [6.2]$$

Depending on the value of the hydraulic depth, the wave velocity may be 10 times higher than the mean water velocity. In reality, the flow will begin to decrease at cross-sections downstream as soon as the leading edge of the translatory wave front arrives, which is much sooner than would be predicted from the average flow velocity in the canal. The total amount of the flow change will be acquired later, at points downstream, because the total flow travels at the average flow velocity in the canal. Once the depressive translatory wave arrives at Checkpoint 2, the water level starts to decrease, but Checkpoint 2 does not have any notice of the true value of this disturbance (2 m^3/s) until 45 min later, once the total amount of the flow change reaches Checkpoint 2.

We can check that the CSI algorithm completely assigns the flow change at P.K. 3 + 750 to the pump station at Checkpoint 2, and this is a logical solution because the water-level variation at Checkpoint 1 is almost null.

The following two remarkable things are noteworthy:

– If the pump flow change value is lower, for instance 50 l/s, and the period of pumping time is much lower, then the CSI algorithm cannot determine the maximum flow change value, due to the hydrograph routing.

– The CSI algorithm assigns the pump flow change introduced at P.K. 3 + 750 entirely to the pump station close to Checkpoint 2. The pump station close to Checkpoint 1 does not pump because the water level measures at Checkpoint 1 are very close to 2 m (water level at Checkpoint 1 in steady state). If the depressive translatory wave[1] reaches Checkpoint 1, then the friction forces greatly attenuate the height of the wave front. For this reason, the water-level changes at Checkpoint 1 are close to 0. This happens because the canal is very steep and the height of the wave front is not sufficiently high to change the water level at Checkpoint 1 significantly. For this reason, this example was tested under the same flow conditions, but by modifying the canal geometry (example 3a); the canal slope was reduced to 0.01% in order to get a higher depressive translatory wave at Checkpoint 1. The results obtained by CSI are shown at Figure 6.9. The results were as expected because the pump flow changes were distributed between the pump stations at Checkpoints 1 and 2, although the pump flow values at Checkpoint 2 were higher than those at Checkpoint 1.

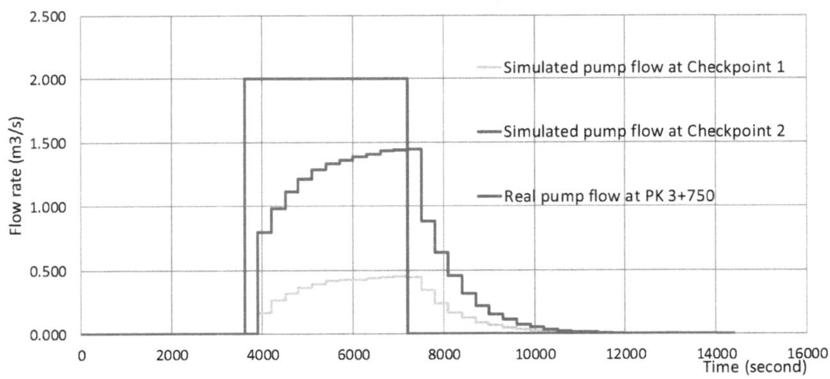

Figure 6.9. *Pump flow values at Checkpoints 1 and 2 during a past time horizon in example 3a. For a color version of the figure, please see www.iste.co.uk/bonet/canals.zip*

1 A translatory wave is a gravity wave that propagates in an open canal and results in the displacement of water particles in a direction parallel to the flow.

However, there is something even more interesting about this scenario.

As we show in Table 6.5, the water volume extracted by the real pump flow at P.K. 3 + 750 (Figure 6.8) is similar to the water volume extracted by the proposed pump flows at Checkpoints 1 and 2 in the third example, with an error of 0.28%. This error value is the same as that in example 3a with a canal slope of 0.01%.

Water volume extracted by the real pump flow hydrograph at P.K. 3 + 750 (m^3)	Water volume extracted by the simulating pump flow hydrograph at Checkpoints 1 and 2 (m^3)	Error (%)
7,200	7,180.4	0.283

Table 6.5. *Water volume error between the simulated and real flow hydrographs in the third example*

The watermaster can determine how much water has been extracted from the canal, and it is also possible to determine the location by using the input and output data of the algorithm.

In this case, the disturbance is introduced in P.K. 3 + 750 at T = 3,600 s. The wave generated by the disturbance in example 3a travels upstream with a velocity (v + c) of 4.21 m/s and downstream with a velocity of 5.20 m/s. Sensors 1 and 2 measure the water level every 5 min (300 s); according to the CSI output data, when the sensor measures the water level at T = 3,900 s, the perturbation is defined at Checkpoints 1 and 2, so that the wave reaches the sensor at Checkpoint 1 just 1 s later than when the sensor measures the water level (T = 3,601 s) or 299 s later at Checkpoint 2 (T = 3,899 s). To obtain the location of the disturbance, four red lines could be drawn in the computational grid of the canal (Figure 6.10): two lines from upstream (v – c) and two lines from downstream (v + c), each defining a wave that arrives 1 s later than when the sensor measures the water level (3,601 s) and 299 s later (3,899 s). The intersections of these lines define an area (A–B–C–D) in the space–time diagram, and every intersection point could be the solution for this disturbance. It is possible to make a good estimation of the pump location, depending on the period of time between water-level measurements.

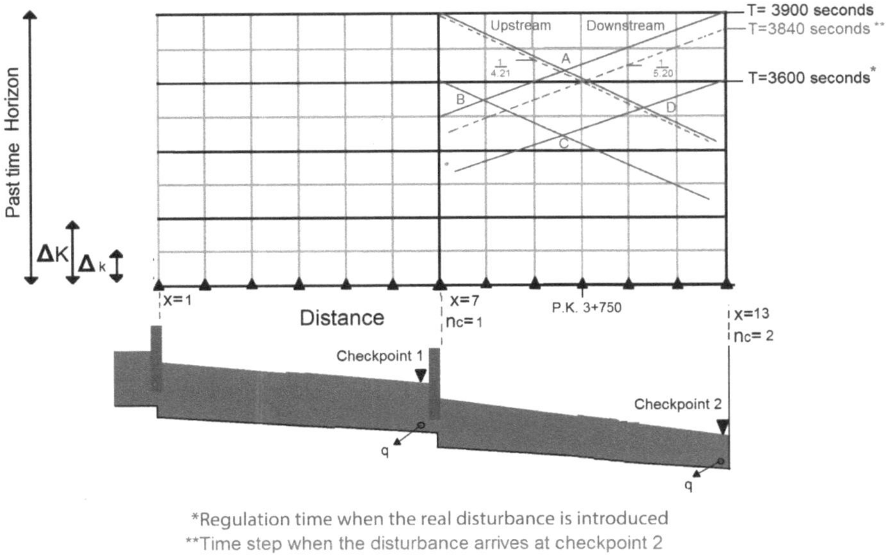

Figure 6.10. *Wave generated by the disturbance traveling upstream and downstream from P.K. 3 + 750. For a color version of the figure, please see www.iste.co.uk/bonet/canals.zip*

In this case, as we know the location, velocity and celerity of the disturbance introduced into the canal, we can determine exactly the time that the wave needs to arrive downstream (Checkpoint 2) and upstream (Checkpoint 1) and easily check if this point is inside of the area (A–B–C–D).

$$Time\ (checkpoint\ 2) = \frac{(X_{P.K.5+000} - X_{P.K.3+750})}{(\text{Velocity} + \text{Celerity})} \approx 240\ seconds$$

$$Time\ (checkpoint\ 1) = \frac{(X_{P.K.3+750} - X_{P.K.2+500})}{(\text{Velocity} + \text{Celerity})} \approx 296\ seconds \qquad [6.3]$$

6.6. Associated errors due to the loss of water-level measurements

Here, we try to set the behavior of the algorithm when the water-level measures at a checkpoint have been lost or simply when the water-level sensor is not working properly. To represent this case, we introduce a new checkpoint (Checkpoint X) and a new pump station with respect to the first example, both located at P.K. 3 + 750. This pump introduces a flow disturbance of 2 m^3/s (at P.K. 3 + 750 for 60 min, from 60th min to 120th min). If the CSI algorithm knows this new pump station at point P.K. 3 + 750, the algorithm assigns a new control structure at node P.K. 3 + 750. However, we do not have water-level measures at this cross-section if the water-level sensor at Checkpoint X (P.K. 3 + 750) is not working.

In that case, the geometry, the discretization of the problem, the initial condition and the disturbance are equal to those in the third example, so we can use the water-level measurements at Checkpoints 1 and 2 of the third example (see section 6.5.3).

The pump flow trajectory simulated by the CSI algorithm is quite different from the pump flow trajectory introduced to the canal, as can be seen in Figure 6.11. Because the disturbance was introduced at P.K. 3 + 750 of the canal, the only pump station that must pump for the CSI algorithm is the pump station at P.K. 3 + 750. However, the pump station at P.K. 5 + 000 is also pumping for the algorithm. The CSI algorithm divides the total flow rate extracted at P.K. 3 + 750. Most of the flow rate is assigned to the pump station at P.K 3 + 750, and the lower flow rate is assigned to the pump station at P.K 5 + 000, as shown in Figure 6.11.

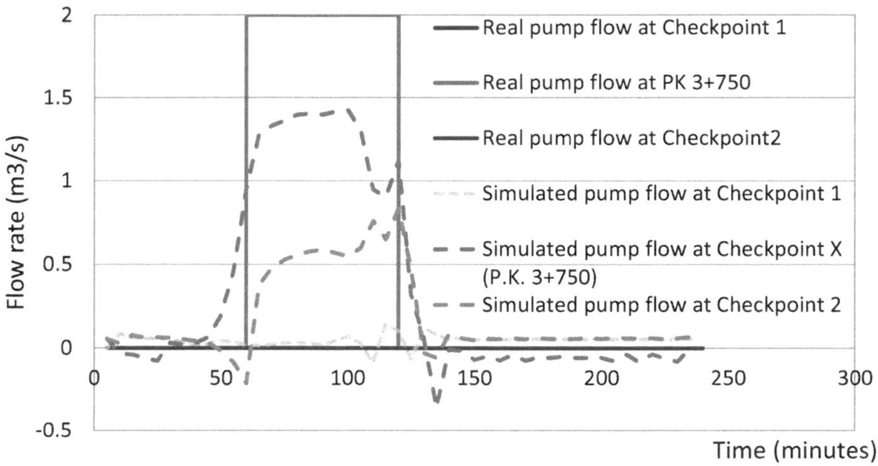

Figure 6.11. *Pump flow at Checkpoints 1 and 2 during a past time horizon. For a color version of the figure, please see www.iste.co.uk/bonet/canals.zip*

The water volume extracted by the real pump flow hydrograph at P.K. 3 + 750 (Figure 6.11) is slightly different from the water volume extracted by the sum of the simulated hydrographs at Checkpoints 1, 2 and X, with an error of 11% (see Table 6.6).

Water volume extracted by the real pump flow hydrograph in P.K. 3 + 750 (m^3)	Water volume extracted by the simulated pump flow hydrograph at Checkpoints 1 and 2 (m^3)	Error (%)
7,200	7,885.78	10.95

Table 6.6. *Water volume error between the simulated and real flow hydrographs*

The expected result in this example would be to assign the pump flow change to the new pump station at P.K. 3 + 750, as this is what really happens. But the question is: if the algorithm does not get water-level measures at P.K. 3 + 750, only at Checkpoints 1 and 2, can the algorithm obtain the real pump flow trajectories at Checkpoints 1, X and 2? The first answer would be no, but we will explain this statement below.

In Figure 6.12, we can observe the water-level measures at the checkpoints introduced to CSI and the water level simulated by the pump flow trajectories calculated by CSI. Both water-level trajectories are similar, but the pump flow trajectories are different. The maximum difference between the measured and simulated water levels at Checkpoint 2 is 2 cm at a time step of 113 min, and the other values are similar with differences of 2–5 mm during the past time horizon. Figure 6.13 only analyzes the results at Checkpoint 2, as it has larger deviation of results.

The CSI algorithm has calculated an extracted flow vector that reproduces the water-level error introduced in the algorithm quite well, but the pump trajectories are not the expected result.

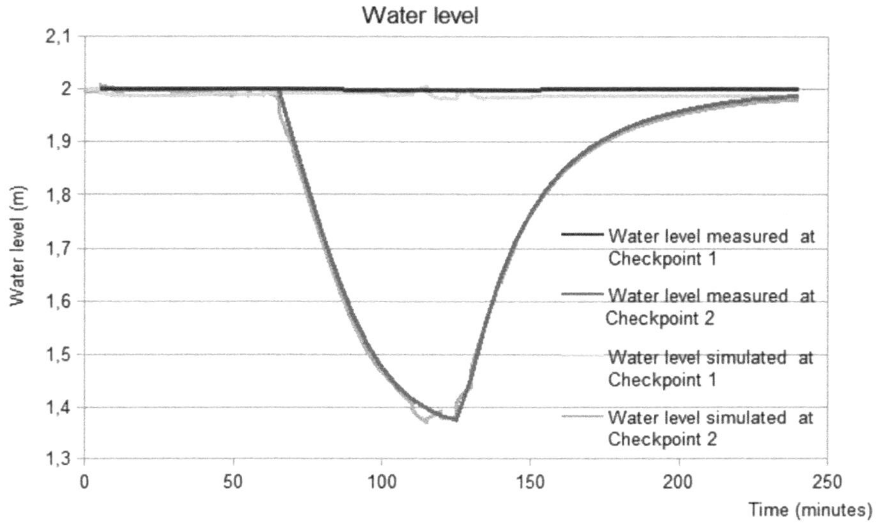

Figure 6.12. *Water levels measured versus simulated by CSI at Checkpoints 1 and 2 during the past time horizon. For a color version of the figure, please see www.iste.co.uk/bonet/canals.zip*

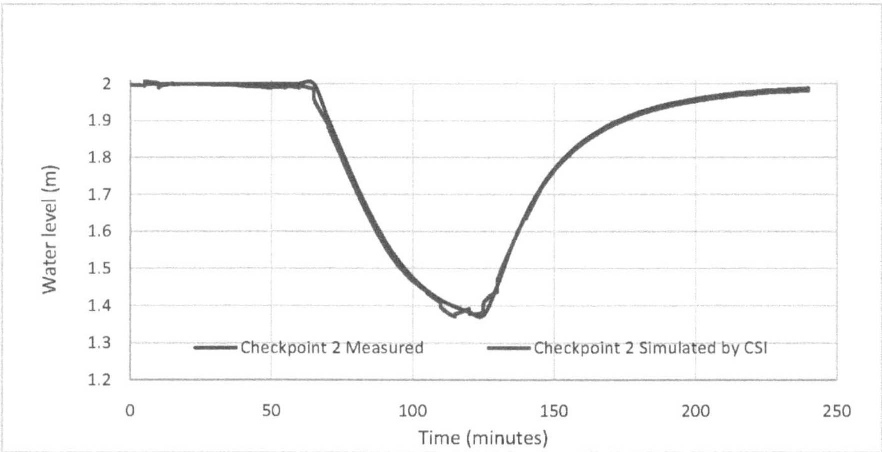

Figure 6.13. *Water levels measured versus simulated by CSI at Checkpoint 2 during the past time horizon. For a color version of the figure, please see www.iste.co.uk/bonet/canals.zip*

6.7. Associated errors due to measurement errors in depth gages

The water level at the checkpoints is measured by sensors which set the water level from an electric signal. Occasionally, this electric signal has an error due to electrical noise or simply due to slight variations of water level at the checkpoint. This error can be significant in some cases, and although the signal is filtered, the water-level measurements provided by the sensor are not fully precise.

In this section, we aim to check the sensitivity of the CSI algorithm to measurement errors. We introduce an error in the water-level measurements and test the first and third examples again with the CSI algorithm. In this way, we introduce three new cases for the first example.

A) We have an error of +2 cm in water-level measurements at Checkpoints 1 and 2 of the first example (flow disturbance), so the water-level function at Checkpoints 1 and 2 (Figure 6.3) is moved.

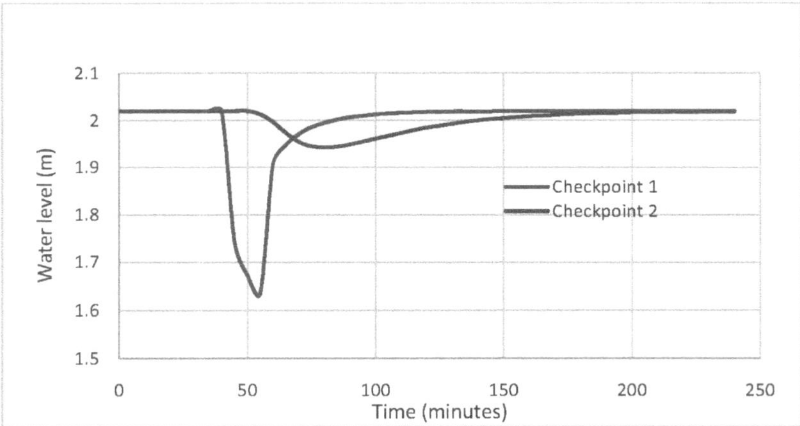

Figure 6.14. *Water levels measured at Checkpoints 1 and 2 during a past time horizon (first example; case: +2 cm). For a color version of the figure, please see www.iste.co.uk/bonet/canals.zip*

B) We introduce an error of −2 cm in water-level measurements at Checkpoints 1 and 2 in the first example (flow disturbance). The water-level function at Checkpoints 1 and 2 (Figure 6.3) is moved.

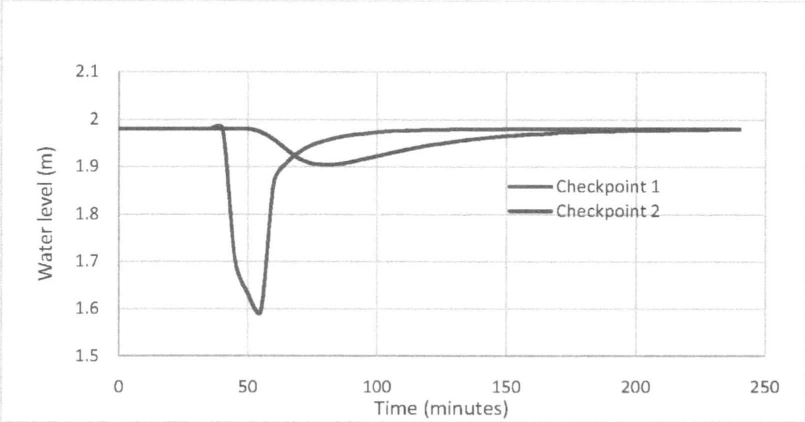

Figure 6.15. *Water levels measured at Checkpoints 1 and 2 during a past time horizon (first example; case: −2 cm). For a color version of the figure, please see www.iste.co.uk/bonet/canals.zip*

C) We introduce an error of +2 cm in the water-level measurements at the first time step, −2 cm at the second time step and so on, at Checkpoints 1 and 2 in the first example. In this sense, the water-level function at Checkpoints 1 and 2 (Figure 6.3) has a sawtooth shape, oscillating with +/−2 cm variation.

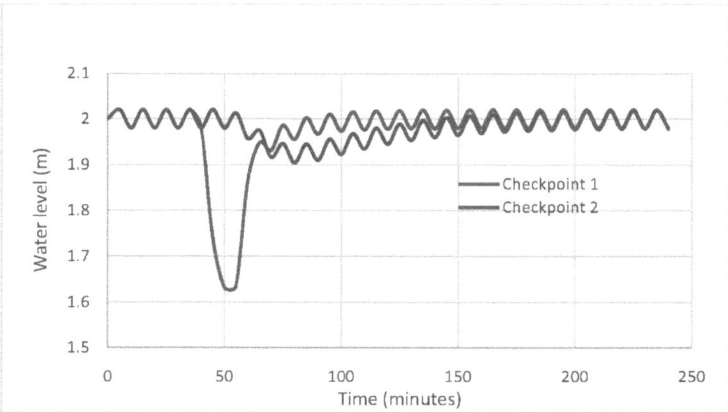

Figure 6.16. *Water levels measured at Checkpoints 1 and 2 during a past time horizon (first example; case: +/−2 cm). For a color version of the figure, please see www.iste.co.uk/bonet/canals.zip*

The results obtained by CSI from these data are shown in Figures 6.17, 6.18 and 6.19.

For all cases, there was an error in the pump flow associated with the error in water-level measurements.

A) The error in the pump flow result is almost constant (Figure 6.17), and the error is easier to define at the end of the test (steady state). In general, we can define that the pump flow error at Checkpoint 1 is approximately –0.084 m^3/s and that at Checkpoint 2 is –0.02 m^3/s. As the depth error increases the real water level by 2 cm, the CSI computes the pump flow necessary to increase the water level at checkpoints by 2 cm. The pump has to introduce a flow of –0.084 m^3/s at Checkpoint 1 and –0.02 m^3/s at Checkpoint 2. The percentage of maximum pump flow error is 4%.

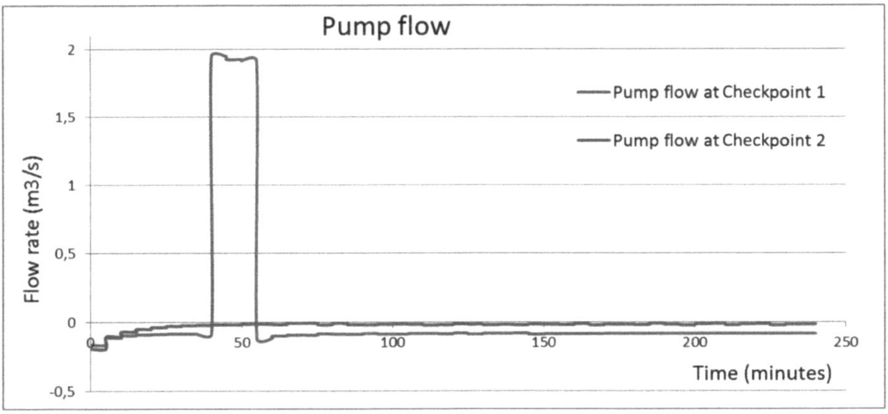

Figure 6.17. *Pump flow at Checkpoints 1 and 2 during the past time horizon (first example; case: +2 cm). For a color version of the figure, please see www.iste.co.uk/bonet/canals.zip*

B) The error in the pump flow result is almost constant in Figure 6.18, and the error is easier to define at the end of the test (steady state). In general, we can see that the pump flow error at Checkpoint 1 is approximately 0.085 m^3/s and that at Checkpoint 2 is 0.016 m^3/s. As the depth error decreases the real water level by 2 cm, the CSI computes the pump flow necessary to reduce the water level at checkpoints by 2 cm. That is, the pump has to extract more flow. The percentage of pump flow error is 4%.

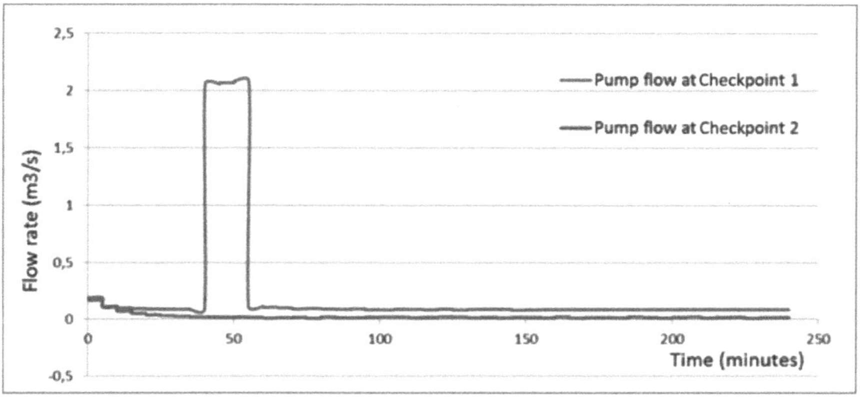

Figure 6.18. *Pump flow at Checkpoints 1 and 2 during the past time horizon (first example; case: −2 cm). For a color version of the figure, please see www.iste.co.uk/bonet/canals.zip*

C) In this case, the error in pump flow changes with the depth error (Figure 6.19) in each regulation time, with an absolute error of 4 cm. For this reason, the pump flow error changes from 0.203 to −0.204 m^3/s at Checkpoint 1 and from 0.242 to −0.245 m^3/s at Checkpoint 2. The percentage of maximum flow error is 12%.

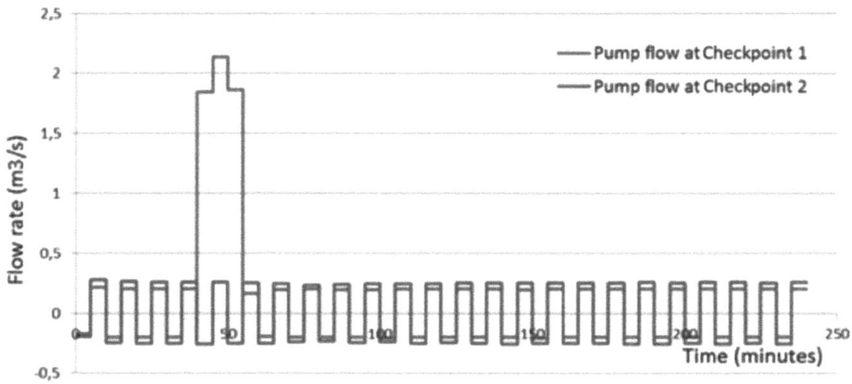

Figure 6.19. *Pump flow at Checkpoints 1 and 2 during the past time horizon (first example; case: +/−2 cm). For a color version of the figure, please see www.iste.co.uk/bonet/canals.zip*

There is a proportional relation between the depth gage error and the pump flow error for this canal.

We have checked the first example by taking into account errors in the water-level measurements. Next, we will test the third example by taking into account an error in the water-level measurements at Checkpoints 1 and 2 of +2 cm. We recall that in the third example the disturbance was introduced at an unknown cross-section for the CSI algorithm, which was not defined as a pump station.

We introduced an error of +2 cm in all water-level measurements at Checkpoints 1 and 2 (see Figure 6.20).

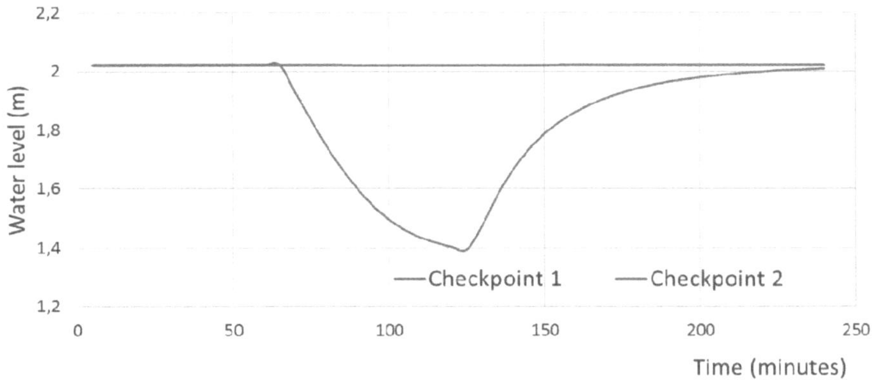

Figure 6.20. *Water level measured at Checkpoints 1 and 2 during the past time horizon (third example; case: +2 cm). For a color version of the figure, please see www.iste.co.uk/bonet/canals.zip*

For all cases, there was an error in the pump flow result associated with the depth gage error in water-level measurements, as in the other examples.

The error in the pump flow result is almost constant compared to the pump flow in the third example, as shown in Figure 6.21. The flow error is similar to the error obtained in the first example with a depth gage error variation of +2 cm. In general, we can define the pump flow error as approximately −0.084 m^3/s at Checkpoint 1 and −0.02 m^3/s at Checkpoint 2. The percentage of maximum pump flow error is 4%. Therefore, the pump flow trajectory obtained in this case is similar to the pump flow trajectory obtained in the third example without depth gage error (Figure 6.8).

As shown in Table 6.7, the water volumes extracted by the real pump flow hydrograph at P.K. 3 + 750 (Figure 6.6) and by the simulated pump flow hydrograph by CSI with depth gage error are quite similar for cases with an error of +2 cm in water-level measurements. The water volume error between the two hydrographs is 6.5% (Table 6.7). The difference in water volume between the two hydrographs is due to the flow rate error of 0.085 m^3/s at Checkpoint 1 and 0.02 m^3/s at Checkpoint 2. Although the flow rate error is low, if the duration of the disturbance is about 95 min (Figure 6.21), then the water volume is not negligible.

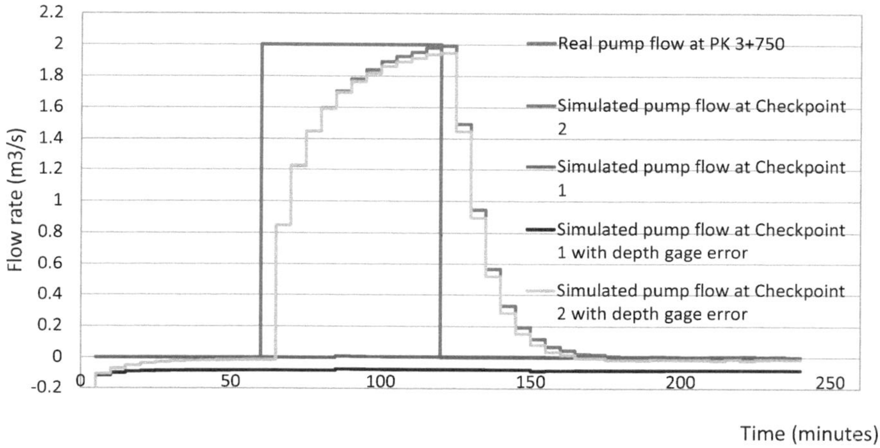

Figure 6.21. *Pump flow at Checkpoints 1 and 2 during the past time horizon (third example; case: +/−2 cm). For a color version of the figure, please see www.iste.co.uk/bonet/canals.zip*

Water volume extracted by the real pump flow hydrograph in P.K. 3 + 750 (m^3)	Water volume extracted by the simulated pump flow hydrograph at Checkpoints 1 and 2 (m^3) in case of an error noise of +2 cm	Error (%)
7,200	6,725.5	6.5

Table 6.7. *Water volume error between the simulated and real pump flow hydrographs from 65th min to 160th min (third example; case: +2 cm)*

6.8. Hydrodynamic canal state

In this section, we show that the CSI algorithm can also define the hydrodynamic state of a canal during a past time horizon and the current canal state, that is, the velocity and water level in each cross-section of the canal. These data are very useful for a feedback controller, which needs to know the hydrodynamic canal state in real time to calculate the optimum gate trajectories.

We have to show that the CSI algorithm is a useful tool for calculating the hydrodynamic canal state, for which we show the hydrodynamic canal state calculated by the CSI algorithm in the first example at 2,700 s (5 min after introducing the disturbance) and at 3,900 s (10 min after removing the disturbance). We can check the water-level measurements at Checkpoints 1 and 2 in Figure 6.22.

Figures 6.23 and 6.24 show water profiles calculated by the CSI algorithm and a computer model (taking into account the real extracted flow), respectively. The error between the real extracted flow and the extracted flow calculated by the CSI algorithm is about 5 l/s in that case, as illustrated in section 6.5.1. Considering the dimensions of the cross-sections of the canal, an error of 5 l/s at a checkpoint represents a water-level variation of about 1 mm in some cross-sections (see Figures 6.23 and 6.24).

In Appendix 1, we show the values of water velocity along the canal obtained by the CSI algorithm and a computer model at 2,700 and 3,900 s.

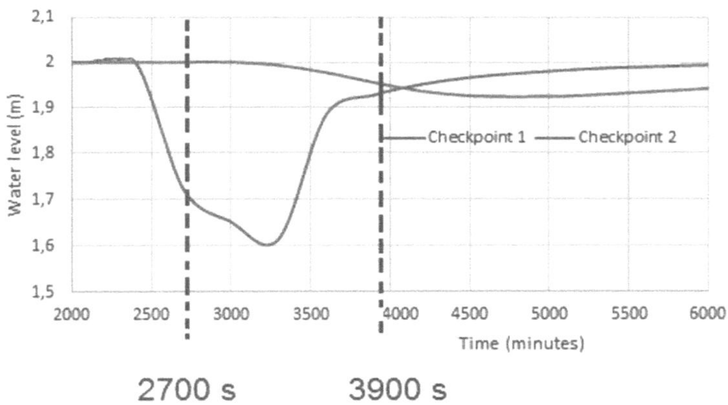

Figure 6.22. *Water-level measurements for the first example (highlighted at 2,700 and 3,900 s). For a color version of the figure, please see www.iste.co.uk/bonet/canals.zip*

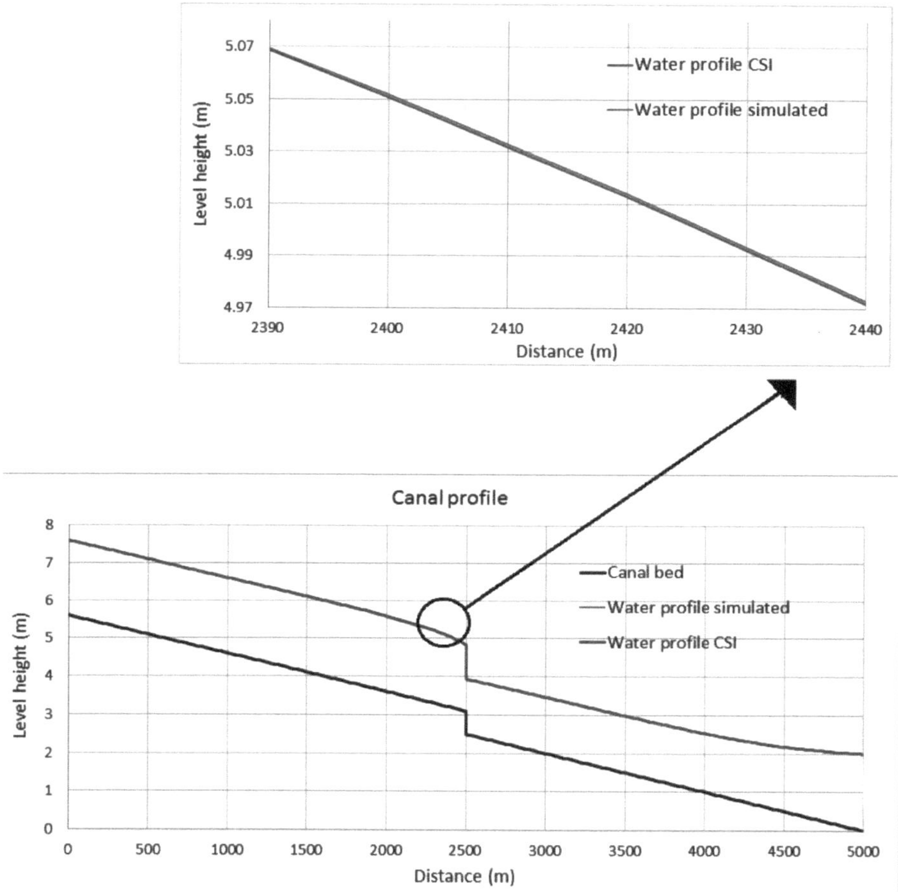

Figure 6.23. *Water profile obtained by a computer model versus water profile obtained by CSI at 2,700 s. For a color version of the figure, please see www.iste.co.uk/bonet/canals.zip*

Both water profiles are similar, as the accuracy of the CSI algorithm in calculating the extracted flow vector is very high. An error of 5 l/s in the extracted flow is equivalent to an error of 1 mm in water level and of 0.001 m/s in water velocity in cross-sections close to the extracted point.

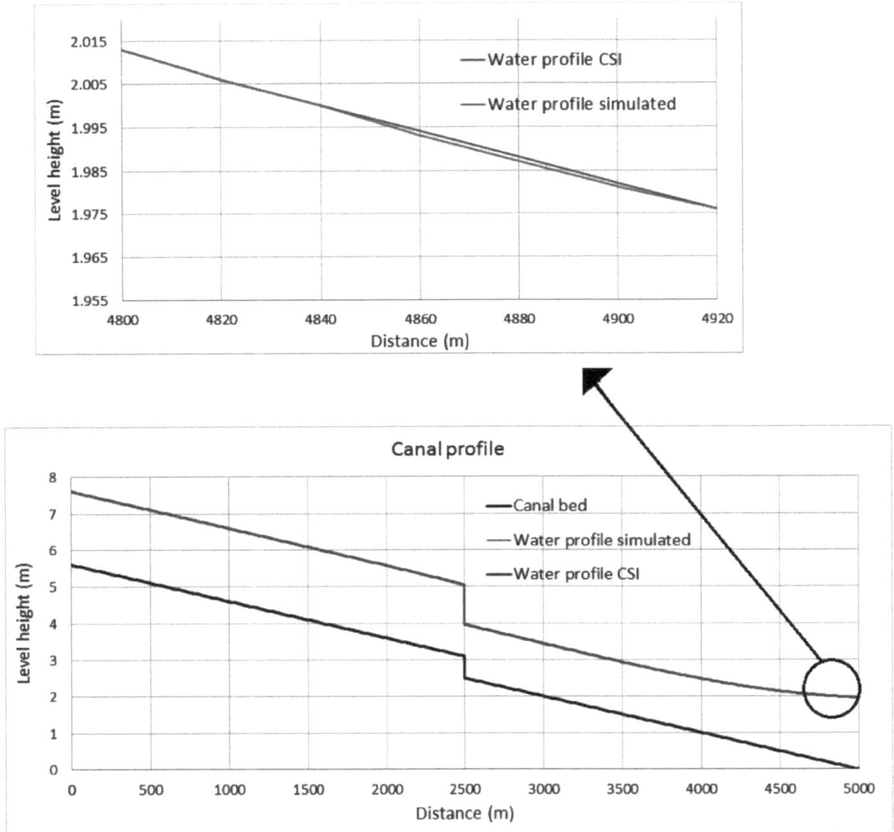

Figure 6.24. *Water profile obtained by a computer model versus water profile obtained by CSI at 3,900 s. For a color version of the figure, please see www.iste.co.uk/bonet/canals.zip*

The second case, at 3,900 s, is equivalent to the other case. Here, the cross-sections with water level errors of about 1 mm are close to the second extraction point, in which the extracted flow error is about 5 l/s.

The hydrodynamic canal states obtained in both cases are similar whether using a computer model operating with the real disturbance or using CSI, which first calculates the disturbance and then the hydrodynamic canal state.

6.9. Conclusion

All these tests have produced excellent results because the algorithm finds a solution to the problem and reproduces the pump flow trajectories correctly in most cases. These examples have allowed us to know the limits of the algorithm, which is very important.

From examples 1 and 2, the real pump flow changes introduced at Checkpoint 1 and the pump flow obtained by CSI are equal, with the maximum error between them lower than 0.005 m^3/s (0.25%) and 0.004 m^3/s (8%), respectively. If the pump flow change is introduced at a node, which CSI recognizes as a pump station and there is a checkpoint associated with this pump station, then the results are good.

The results obtained in example 3 are not so good but still valuable. In example 3, although it is impossible to estimate correctly the pump flow hydrograph, it is possible to estimate how much water is extracted from the canal with a maximum error of 0.28%. It is also possible to locate the cross-section where the pump extracts the flow rate quite well, with more or less accuracy depending on the time frequency between measurements and the value of the pump flow change. This is not an unusual scenario in a real canal because we cannot always count on having checkpoints close to the pump which is operating, as the disturbances are usually introduced without notice and without the knowledge of the watermaster.

If a pump extracts water from an unknown cross-section of the canal and the algorithm has not assigned any checkpoint in this localization, then the CSI algorithm can obtain the total volume of water extracted by the pump.

For errors associated with the loss of water-level measurements at checkpoints, we show that it is not a good idea to have more pump stations or variables than checkpoints or measurement points. If there are more pump stations than measurement points, then there will be multiple solutions or pump flow trajectories for reaching similar water-level measures at the checkpoints. Instead, there is only one solution when we know which pump extracts water and which does not. The pump flow results are striking in example 3, when the algorithm knows that the only operative pump station is the one at P.K. 3 + 750.

In case of loss of accuracy in water-level sensors, we have to be careful with the data obtained from the gages. In those cases, the CSI algorithm has obtained good results, taking into account an error of 2 cm in example 1, and the pump flow result only deviates by a maximum of 0.085 m^3/s, from 2 to 2.085 m^3/s. We have also repeated the third example, considering an error of 2 cm. The pump flow trajectories calculated by the CSI algorithm with and without error are similar to the real pump flow trajectory, and the water volume error is not significant with respect to the real extracted water volume (6.5%).

We have also calculated the hydrodynamic canal state at two time instances, 2,700 and 3,900 s, in the first example. The accuracy of the CSI algorithm in calculating the extracted flow vector is essential for accurately calculating the hydrodynamic canal state. Using the algorithm, we calculate the hydrodynamic canal state from the scheduled demands, gate trajectories, initial conditions and the extracted flow vector. The hydrodynamic canal states obtained by a computer model using the real disturbances and using the CSI algorithm (previously computing the disturbance) are very similar.

The algorithm is as good as the results show: it works well with small disturbances as well as in cases where the flow rate pumping is not located in a control structure.

6.10. Appendix

Time step (s)	Distance (m)	Water velocity by CSI (m/s)	Water velocity by the model (m/s)	Time step (s)	Distance (m)	Water velocity by CSI (m/s)	Water velocity by the model (m/s)
2.700	0	1.245	1.245	3900	0	1.245	1.245
2.700	20	1.245	1.245	3901	20	1.245	1.245
2.700	40	1.245	1.245	3902	40	1.245	1.245
2.700	60	1.245	1.245	3903	60	1.245	1.245
2.700	80	1.245	1.245	3904	80	1.245	1.245
2.700	100	1.245	1.245	3905	100	1.245	1.245
2.700	120	1.245	1.245	3906	120	1.245	1.245
2.700	140	1.245	1.245	3907	140	1.245	1.245
2.700	160	1.245	1.245	3908	160	1.245	1.245
2.700	180	1.245	1.245	3909	180	1.245	1.245

Time step (s)	Distance (m)	Water velocity by CSI (m/s)	Water velocity by the model (m/s)	Time step (s)	Distance (m)	Water velocity by CSI (m/s)	Water velocity by the model (m/s)
2.700	200	1.245	1.245	3910	200	1.245	1.245
2.700	220	1.245	1.245	3911	220	1.245	1.245
2.700	240	1.245	1.245	3912	240	1.245	1.245
2.700	260	1.245	1.245	3913	260	1.245	1.245
2.700	280	1.245	1.245	3914	280	1.245	1.245
2.700	300	1.245	1.245	3915	300	1.245	1.245
2.700	320	1.245	1.245	3916	320	1.245	1.245
2.700	340	1.245	1.245	3917	340	1.245	1.245
2.700	360	1.245	1.245	3918	360	1.245	1.245
2.700	380	1.245	1.245	3919	380	1.245	1.245
2.700	400	1.245	1.245	3920	400	1.245	1.245
2.700	420	1.245	1.245	3921	420	1.245	1.245
2.700	440	1.245	1.245	3922	440	1.245	1.245
2.700	460	1.245	1.245	3923	460	1.245	1.245
2.700	480	1.245	1.245	3924	480	1.245	1.245
2.700	500	1.245	1.245	3925	500	1.245	1.245
2.700	520	1.245	1.245	3926	520	1.245	1.245
2.700	540	1.245	1.245	3927	540	1.245	1.245
2.700	560	1.245	1.245	3928	560	1.246	1.246
2.700	580	1.245	1.245	3929	580	1.246	1.246
2.700	600	1.245	1.245	3930	600	1.246	1.246
2.700	620	1.245	1.245	3931	620	1.246	1.246
2.700	640	1.245	1.245	3932	640	1.246	1.246

Time step (s)	Distance (m)	Water velocity by CSI (m/s)	Water velocity by the model (m/s)	Time step (s)	Distance (m)	Water velocity by CSI (m/s)	Water velocity by the model (m/s)
2.700	660	1.245	1.245	3933	660	1.246	1.246
2.700	680	1.245	1.245	3934	680	1.246	1.246
2.700	700	1.245	1.245	3935	700	1.246	1.246
2.700	720	1.245	1.245	3936	720	1.246	1.246
2.700	740	1.245	1.245	3937	740	1.246	1.246
2.700	760	1.245	1.245	3938	760	1.246	1.246
2.700	780	1.245	1.245	3939	780	1.246	1.246
2.700	800	1.245	1.245	3940	800	1.246	1.246
2.700	820	1.245	1.245	3941	820	1.247	1.247
2.700	840	1.245	1.245	3942	840	1.247	1.247
2.700	860	1.245	1.245	3943	860	1.247	1.247
2.700	880	1.245	1.245	3944	880	1.247	1.247
2.700	900	1.245	1.245	3945	900	1.247	1.247
2.700	920	1.245	1.245	3946	920	1.247	1.247
2.700	940	1.245	1.245	3947	940	1.247	1.247
2.700	960	1.245	1.245	3948	960	1.248	1.248
2.700	980	1.245	1.245	3949	980	1.248	1.248
2.700	1000	1.245	1.245	3950	1000	1.248	1.248

Time step (s)	Distance (m)	Water velocity by CSI (m/s)	Water velocity by the model (m/s)	Time step (s)	Distance (m)	Water velocity by CSI (m/s)	Water velocity by the model (m/s)
2.700	1020	1.245	1.245	3951	1020	1.248	1.248
2.700	1040	1.245	1.245	3952	1040	1.248	1.248
2.700	1060	1.245	1.245	3953	1060	1.249	1.249
2.700	1080	1.245	1.245	3954	1080	1.249	1.249
2.700	1100	1.245	1.245	3955	1100	1.249	1.249
2.700	1120	1.245	1.245	3956	1120	1.25	1.25
2.700	1140	1.245	1.245	3957	1140	1.25	1.25
2.700	1160	1.245	1.245	3958	1160	1.25	1.25
2.700	1180	1.245	1.245	3959	1180	1.25	1.25
2.700	1200	1.245	1.245	3960	1200	1.251	1.251
2.700	1220	1.245	1.245	3961	1220	1.251	1.251
2.700	1240	1.245	1.245	3962	1240	1.252	1.252
2.700	1260	1.245	1.245	3963	1260	1.252	1.252
2.700	1280	1.245	1.245	3964	1280	1.252	1.252
2.700	1300	1.245	1.245	3965	1300	1.253	1.253
2.700	1320	1.245	1.245	3966	1320	1.253	1.253
2.700	1340	1.245	1.245	3967	1340	1.253	1.253
2.700	1360	1.245	1.245	3968	1360	1.254	1.254
2.700	1380	1.245	1.245	3969	1380	1.254	1.254
2.700	1400	1.245	1.245	3970	1400	1.254	1.254

Time step (s)	Distance (m)	Water velocity by CSI (m/s)	Water velocity by the model (m/s)	Time step (s)	Distance (m)	Water velocity by CSI (m/s)	Water velocity by the model (m/s)
2.700	1420	1.245	1.245	3971	1420	1.255	1.255
2.700	1440	1.245	1.245	3972	1440	1.255	1.255
2.700	1460	1.245	1.245	3973	1460	1.255	1.255
2.700	1480	1.245	1.245	3974	1480	1.256	1.256
2.700	1500	1.245	1.245	3975	1500	1.256	1.256
2.700	1520	1.245	1.245	3976	1520	1.257	1.257
2.700	1540	1.245	1.245	3977	1540	1.257	1.257
2.700	1560	1.245	1.245	3978	1560	1.258	1.258
2.700	1580	1.245	1.245	3979	1580	1.258	1.258
2.700	1600	1.245	1.245	3980	1600	1.259	1.259
2.700	1620	1.245	1.245	3981	1620	1.259	1.259
2.700	1640	1.245	1.245	3982	1640	1.26	1.26
2.700	1660	1.245	1.245	3983	1660	1.26	1.26
2.700	1680	1.245	1.245	3984	1680	1.261	1.261
2.700	1700	1.245	1.245	3985	1700	1.261	1.261
2.700	1720	1.245	1.245	3986	1720	1.262	1.262
2.700	1740	1.246	1.246	3987	1740	1.262	1.262
2.700	1760	1.246	1.246	3988	1760	1.263	1.263

Time step (s)	Distance (m)	Water velocity by CSI (m/s)	Water velocity by the model (m/s)	Time step (s)	Distance (m)	Water velocity by CSI (m/s)	Water velocity by the model (m/s)
2.700	1780	1.246	1.246	3989	1780	1.263	1.263
2.700	1800	1.246	1.246	3990	1800	1.264	1.264
2.700	1820	1.247	1.247	3991	1820	1.264	1.264
2.700	1840	1.249	1.249	3992	1840	1.265	1.265
2.700	1860	1.251	1.251	3993	1860	1.266	1.266
2.700	1880	1.253	1.253	3994	1880	1.266	1.266
2.700	1900	1.257	1.257	3995	1900	1.267	1.267
2.700	1920	1.262	1.262	3996	1920	1.268	1.268
2.700	1940	1.267	1.267	3997	1940	1.268	1.268
2.700	1960	1.273	1.273	3998	1960	1.269	1.269
2.700	1980	1.278	1.278	3999	1980	1.27	1.27
2.700	2000	1.283	1.283	4000	2000	1.27	1.27
2.700	2020	1.288	1.288	4001	2020	1.271	1.271
2.700	2040	1.293	1.292	4002	2040	1.272	1.272
2.700	2060	1.297	1.297	4003	2060	1.272	1.272
2.700	2080	1.302	1.302	4004	2080	1.273	1.273
2.700	2100	1.308	1.308	4005	2100	1.274	1.274
2.700	2120	1.314	1.314	4006	2120	1.275	1.275
2.700	2140	1.321	1.321	4007	2140	1.276	1.276
2.700	2160	1.329	1.329	4008	2160	1.276	1.276

Time step (s)	Distance (m)	Water velocity by CSI (m/s)	Water velocity by the model (m/s)	Time step (s)	Distance (m)	Water velocity by CSI (m/s)	Water velocity by the model (m/s)
2.700	2180	1.338	1.337	4009	2180	1.277	1.277
2.700	2200	1.347	1.346	4010	2200	1.278	1.278
2.700	2220	1.356	1.356	4011	2220	1.279	1.279
2.700	2240	1.367	1.367	4012	2240	1.28	1.28
2.700	2260	1.379	1.379	4013	2260	1.281	1.281
2.700	2280	1.392	1.392	4014	2280	1.282	1.282
2.700	2300	1.407	1.406	4015	2300	1.283	1.283
2.700	2320	1.423	1.422	4016	2320	1.284	1.284
2.700	2340	1.441	1.44	4017	2340	1.285	1.285
2.700	2360	1.461	1.46	4018	2360	1.286	1.286
2.700	2380	1.484	1.483	4019	2380	1.287	1.287
2.700	2400	1.51	1.509	4020	2400	1.288	1.288
2.700	2420	1.54	1.539	4021	2420	1.289	1.289
2.700	2440	1.575	1.575	4022	2440	1.29	1.29
2.700	2460	1.618	1.617	4023	2460	1.291	1.291
2.700	2480	1.67	1.669	4024	2480	1.293	1.292
2.700	2500	1.736	1.735	4025	2500	1.294	1.294
2.700	2500	0.99	0.99	4026	2500	1.052	1.052

Time step (s)	Distance (m)	Water velocity by CSI (m/s)	Water velocity by the model (m/s)	Time step (s)	Distance (m)	Water velocity by CSI (m/s)	Water velocity by the model (m/s)
2.700	2520	0.991	0.991	4027	2520	1.052	1.052
2.700	2540	0.991	0.992	4028	2540	1.052	1.052
2.700	2560	0.992	0.992	4029	2560	1.052	1.052
2.700	2580	0.993	0.993	4030	2580	1.052	1.052
2.700	2600	0.994	0.994	4031	2600	1.051	1.051
2.700	2620	0.995	0.995	4032	2620	1.051	1.051
2.700	2640	0.996	0.996	4033	2640	1.051	1.051
2.700	2660	0.996	0.997	4034	2660	1.051	1.051
2.700	2680	0.997	0.997	4035	2680	1.051	1.051
2.700	2700	0.998	0.998	4036	2700	1.05	1.05
2.700	2720	0.999	0.999	4037	2720	1.05	1.05
2.700	2740	1	1	4038	2740	1.05	1.05
2.700	2760	1.001	1.001	4039	2760	1.049	1.049
2.700	2780	1.002	1.002	4040	2780	1.049	1.049
2.700	2800	1.003	1.003	4041	2800	1.049	1.049
2.700	2820	1.004	1.004	4042	2820	1.048	1.048
2.700	2840	1.005	1.005	4043	2840	1.048	1.048
2.700	2860	1.006	1.006	4044	2860	1.048	1.048
2.700	2880	1.007	1.007	4045	2880	1.047	1.047
2.700	2900	1.008	1.008	4046	2900	1.047	1.047

Time step (s)	Distance (m)	Water velocity by CSI (m/s)	Water velocity by the model (m/s)	Time step (s)	Distance (m)	Water velocity by CSI (m/s)	Water velocity by the model (m/s)
2.700	2920	1.009	1.009	4047	2920	1.046	1.046
2.700	2940	1.01	1.01	4048	2940	1.046	1.046
2.700	2960	1.011	1.011	4049	2960	1.045	1.045
2.700	2980	1.012	1.012	4050	2980	1.045	1.045
2.700	3000	1.013	1.013	4051	3000	1.044	1.044
2.700	3020	1.014	1.014	4052	3020	1.043	1.043
2.700	3040	1.015	1.015	4053	3040	1.043	1.043
2.700	3060	1.015	1.015	4054	3060	1.042	1.042
2.700	3080	1.016	1.016	4055	3080	1.042	1.042
2.700	3100	1.017	1.017	4056	3100	1.041	1.041
2.700	3120	1.018	1.018	4057	3120	1.04	1.04
2.700	3140	1.019	1.019	4058	3140	1.039	1.039
2.700	3160	1.02	1.02	4059	3160	1.039	1.039
2.700	3180	1.02	1.02	4060	3180	1.038	1.038
2.700	3200	1.021	1.021	4061	3200	1.037	1.037
2.700	3220	1.022	1.022	4062	3220	1.036	1.036
2.700	3240	1.023	1.023	4063	3240	1.035	1.035
2.700	3260	1.023	1.023	4064	3260	1.034	1.034

Time step (s)	Distance (m)	Water velocity by CSI (m/s)	Water velocity by the model (m/s)	Time step (s)	Distance (m)	Water velocity by CSI (m/s)	Water velocity by the model (m/s)
2.700	3280	1.024	1.024	4065	3280	1.034	1.034
2.700	3300	1.024	1.024	4066	3300	1.033	1.033
2.700	3320	1.025	1.025	4067	3320	1.032	1.032
2.700	3340	1.025	1.025	4068	3340	1.031	1.03
2.700	3360	1.026	1.026	4069	3360	1.029	1.029
2.700	3380	1.026	1.026	4070	3380	1.028	1.028
2.700	3400	1.026	1.026	4071	3400	1.027	1.027
2.700	3420	1.027	1.027	4072	3420	1.026	1.026
2.700	3440	1.027	1.027	4073	3440	1.025	1.025
2.700	3460	1.027	1.027	4074	3460	1.023	1.023
2.700	3480	1.027	1.027	4075	3480	1.022	1.022
2.700	3500	1.027	1.027	4076	3500	1.021	1.021
2.700	3520	1.027	1.027	4077	3520	1.019	1.019
2.700	3540	1.026	1.026	4078	3540	1.018	1.018
2.700	3560	1.026	1.026	4079	3560	1.016	1.016
2.700	3580	1.026	1.026	4080	3580	1.015	1.015
2.700	3600	1.026	1.026	4081	3600	1.013	1.013
2.700	3620	1.026	1.026	4082	3620	1.011	1.011
2.700	3640	1.025	1.025	4083	3640	1.01	1.01
2.700	3660	1.025	1.025	4084	3660	1.008	1.008

Time step (s)	Distance (m)	Water velocity by CSI (m/s)	Water velocity by the model (m/s)	Time step (s)	Distance (m)	Water velocity by CSI (m/s)	Water velocity by the model (m/s)
2.700	3680	1.025	1.025	4085	3680	1.006	1.006
2.700	3700	1.024	1.024	4086	3700	1.004	1.004
2.700	3720	1.023	1.023	4087	3720	1.002	1.002
2.700	3740	1.022	1.022	4088	3740	1	1
2.700	3760	1.02	1.02	4089	3760	0.997	0.997
2.700	3780	1.018	1.018	4090	3780	0.995	0.995
2.700	3800	1.016	1.016	4091	3800	0.993	0.993
2.700	3820	1.014	1.014	4092	3820	0.99	0.99
2.700	3840	1.012	1.012	4093	3840	0.987	0.987
2.700	3860	1.01	1.01	4094	3860	0.985	0.985
2.700	3880	1.008	1.008	4095	3880	0.982	0.982
2.700	3900	1.005	1.005	4096	3900	0.979	0.979
2.700	3920	1.002	1.002	4097	3920	0.976	0.976
2.700	3940	1	1	4098	3940	0.973	0.972
2.700	3960	0.997	0.997	4099	3960	0.969	0.969
2.700	3980	0.994	0.994	4100	3980	0.966	0.966
2.700	4000	0.99	0.99	4101	4000	0.962	0.962
2.700	4020	0.987	0.987	4102	4020	0.958	0.958

Time step (s)	Distance (m)	Water velocity by CSI (m/s)	Water velocity by the model (m/s)	Time step (s)	Distance (m)	Water velocity by CSI (m/s)	Water velocity by the model (m/s)
2.700	4040	0.983	0.983	4103	4040	0.954	0.954
2.700	4060	0.979	0.979	4104	4060	0.95	0.95
2.700	4080	0.975	0.975	4105	4080	0.946	0.946
2.700	4100	0.971	0.971	4106	4100	0.942	0.942
2.700	4120	0.966	0.966	4107	4120	0.937	0.937
2.700	4140	0.962	0.962	4108	4140	0.933	0.932
2.700	4160	0.957	0.957	4109	4160	0.928	0.928
2.700	4180	0.952	0.952	4110	4180	0.923	0.923
2.700	4200	0.947	0.947	4111	4200	0.918	0.918
2.700	4220	0.941	0.941	4112	4220	0.912	0.912
2.700	4240	0.936	0.936	4113	4240	0.907	0.907
2.700	4260	0.93	0.93	4114	4260	0.901	0.901
2.700	4280	0.924	0.924	4115	4280	0.895	0.895
2.700	4300	0.917	0.917	4116	4300	0.889	0.889
2.700	4320	0.911	0.911	4117	4320	0.883	0.883
2.700	4340	0.904	0.904	4118	4340	0.877	0.877
2.700	4360	0.897	0.897	4119	4360	0.871	0.871
2.700	4380	0.89	0.89	4120	4380	0.864	0.864
2.700	4400	0.883	0.883	4121	4400	0.858	0.858
2.700	4420	0.875	0.875	4122	4420	0.851	0.851

Time step (s)	Distance (m)	Water velocity by CSI (m/s)	Water velocity by the model (m/s)	Time step (s)	Distance (m)	Water velocity by CSI (m/s)	Water velocity by the model (m/s)
2.700	4440	0.868	0.868	4123	4440	0.844	0.844
2.700	4460	0.86	0.86	4124	4460	0.837	0.837
2.700	4480	0.852	0.852	4125	4480	0.83	0.83
2.700	4500	0.844	0.844	4126	4500	0.823	0.823
2.700	4520	0.836	0.836	4127	4520	0.816	0.816
2.700	4540	0.828	0.828	4128	4540	0.808	0.808
2.700	4560	0.819	0.819	4129	4560	0.801	0.801
2.700	4580	0.811	0.811	4130	4580	0.793	0.794
2.700	4600	0.802	0.802	4131	4600	0.786	0.786
2.700	4620	0.793	0.793	4132	4620	0.778	0.779
2.700	4640	0.785	0.785	4133	4640	0.771	0.771
2.700	4660	0.776	0.776	4134	4660	0.763	0.763
2.700	4680	0.767	0.767	4135	4680	0.755	0.756
2.700	4700	0.758	0.758	4136	4700	0.748	0.748
2.700	4720	0.749	0.749	4137	4720	0.74	0.74
2.700	4740	0.74	0.74	4138	4740	0.732	0.733
2.700	4760	0.731	0.731	4139	4760	0.724	0.725
2.700	4780	0.722	0.722	4140	4780	0.717	0.717
2.700	4800	0.713	0.713	4141	4800	0.709	0.709
2.700	4820	0.704	0.704	4142	4820	0.701	0.702

Time step (s)	Distance (m)	Water velocity by CSI (m/s)	Water velocity by the model (m/s)	Time step (s)	Distance (m)	Water velocity by CSI (m/s)	Water velocity by the model (m/s)
2.700	4840	0.695	0.695	4143	4840	0.694	0.694
2.700	4860	0.686	0.686	4144	4860	0.686	0.687
2.700	4880	0.677	0.677	4145	4880	0.679	0.679
2.700	4900	0.668	0.668	4146	4900	0.671	0.672
2.700	4920	0.659	0.659	4147	4920	0.664	0.664
2.700	4940	0.651	0.651	4148	4940	0.657	0.657
2.700	4960	0.642	0.642	4149	4960	0.649	0.65
2.700	4980	0.634	0.634	4150	4980	0.642	0.643
2.700	5000	0.625	0.625	4151	5000	0.635	0.636

Table 6.8. *Water velocity obtained by the computer model (real disturbance) versus water velocity obtained by the CSI algorithm along the canal in the first example at 2,700 and 3,900 s*

7

CSI Tests on the Experimental Canal PAC-UPC

In this chapter, we test the CSI algorithm in a real canal. We aim to prove that the CSI algorithm is used to accurately calculate a disturbance introduced into a canal; therefore, we performed some tests in a laboratory canal to verify the results obtained by CSI described in Chapter 6.

Canal PAC-UPC is a laboratory canal specially designed to develop basic and applied research in irrigation canal control areas and in all subjacent areas like irrigation canal instrumentation, irrigation canal modeling and water measurements.

The canal is located in the Laboratory of Physical Models inside the North Campus of the UPC. This laboratory has a surface area of 2,000 m^2.

Most of the design, acquisition and final implementation of the instrumentation and of the motorization of the canal were carried out by Carlos Alberto Sepúlveda Toepfer, who described all of this work in his thesis (see Sepúlveda [SEP 07]).

7.1. General description

The original idea was to build a canal that could exhibit notorious transport delays in order to resemble real irrigation canal control problems. Therefore, it was decided to construct a long zero-slope rectangular cross-section canal, but due to space constraints inside the laboratory, the canal was designed with a serpentine shape to achieve the maximum canal length in the available space. With this particular design, a canal with length 220 m, width 44 cm and depth 1 m was developed. A detailed scheme of this laboratory canal is presented in Figures 7.1 and 7.2.

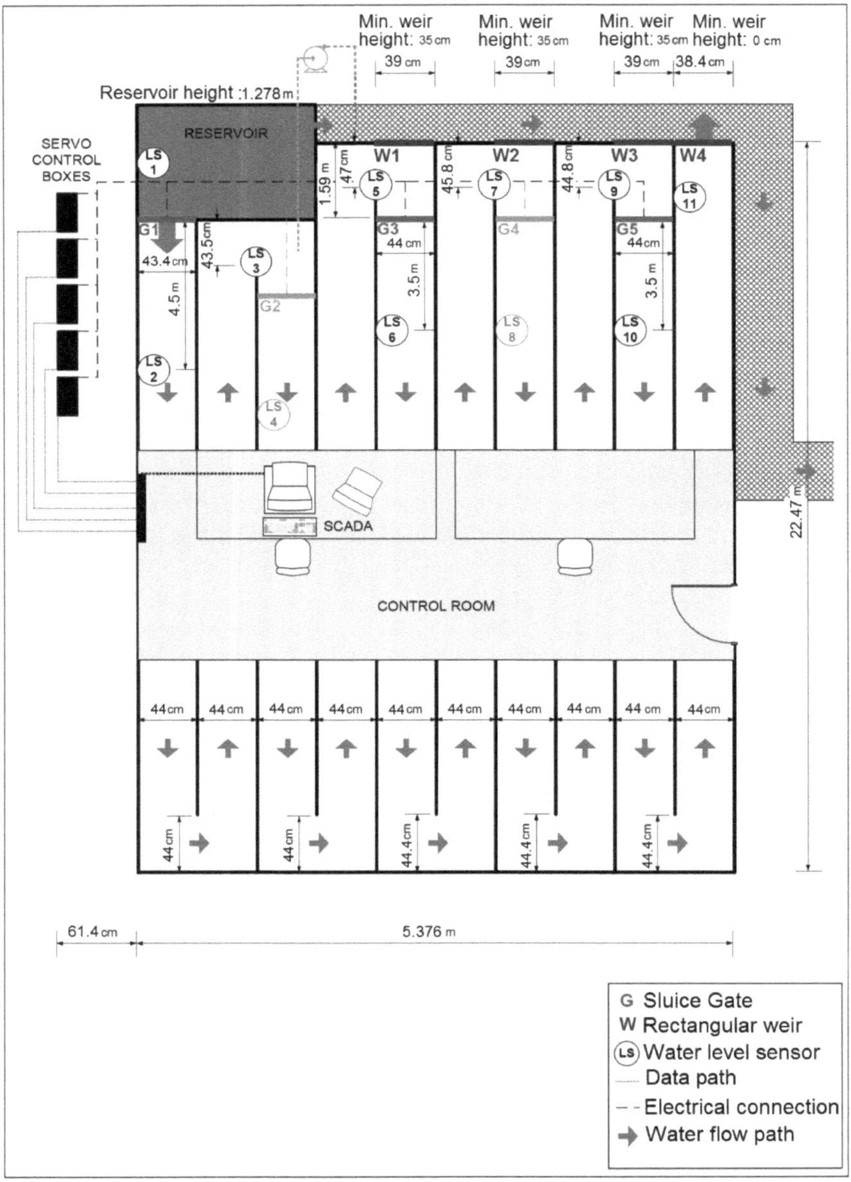

Figure 7.1. *Schematic diagram of the top view of the Canal PAC-UPC. For a color version of the figure, please see www.iste.co.uk/bonet/canals.zip*

The canal is supplied by a small reservoir at its head. The objective of this element is to provide the canal with enough, that is, virtually unlimited, water.

The canal takes water from this reservoir through Gate1 (G1), which is normally under submerged conditions. This gate can regulate the canal inflow by adjusting the gate opening. The unused water is recirculated to the laboratory pumping system.

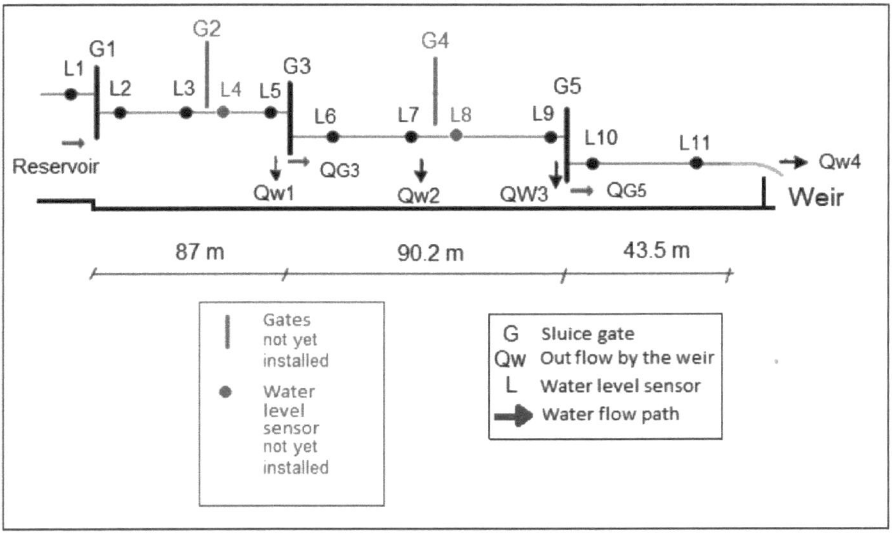

Figure 7.2. *Schematic diagram of the profile view of the Canal PAC-UPC. For a color version of the figure, please see www.iste.co.uk/bonet/canals.zip*

As illustrated in Figures 7.1 and 7.2, the canal currently has:

– one head reservoir;

– three vertical sluice gates;

– four rectangular weirs;

– nine water level sensors; and

– one control room.

Under the present condition, it is possible to arrange the canal with several pool configurations, for example, a canal with only one very long pool and a canal with one long pool and one short pool.

Figure 7.3 shows a general view of the canal.

Figure 7.3. *General view of the Canal PAC-UPC*

Canal geometry test

In this test, the canal with the following geometric configuration is used: the sluice gate 1 is the only sluice gate operating in this test because gates 3 and 5 are out of water. Only weirs W2 and W4 are operating, and the algorithm only uses the water level measured at sensors L1, L6, L10 and L11; however, we will also use the data of the sensor L7 to check some results. The canal geometry configuration used in CSI is shown in Table 7.1 and Figure 7.4.

Pool	Numerical node upstream	Numerical node downstream	Pool length (m)	Canal depth (m)	Manning's coefficient (n)	Width (m)
1 (from G1 to G2)	1	43	42	1.00	0.016	0.44
2 (from G2 to G3)	44	89	45	1.00	0.016	0.44
3 (from G3 to G4)	90	135	45	1.00	0.016	0.44
4 (from G4 to G5)	136	181	45	1.00	0.015	0.44
5 (from G5 to W4)	182	225	43	1.00	0.016	0.44

Table 7.1. *Features of the canal pools*

We could not use all the sensors because some of them were out of order and others were located very close to the canal bend. In every bend of the canal, the water surface has variations of ±1 cm due to the local flow pattern.

The CSI algorithm is very sensitive to water-level variations at the checkpoints; furthermore, these water-level variations are equivalent to the water-level changes introduced to the system by unknown discharge. In that case, the algorithm cannot recognize if the changes in water-level at the checkpoints are introduced by unknown discharge. In such circumstances, the algorithm calculates an extracted flow change considering the absolute value of the water-level changes, and therefore, the extracted flow simulated is different from the real unknown discharge introduced into the canal.

The signal can be filtered, but if the water-level changes due to unknown disturbances are similar to the noise generated by the canal bend, then it is impossible to distinguish them.

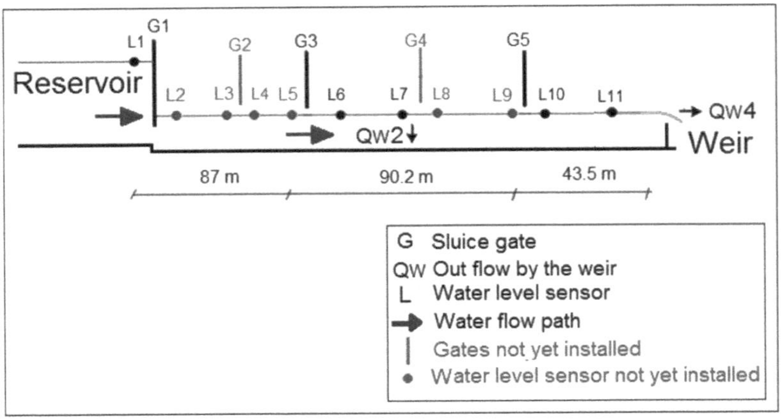

Figure 7.4. *Profile view for the geometric configuration of the Canal PAC-UPC in the test. For a color version of the figure, please see www.iste.co.uk/bonet/canals.zip*

The CSI algorithm does not simulate the weirs of the canal although it could be possible; that is, it only obtains the outflows. The advantages of working with the flow are that the control action variable directly reflects the flow rate and that the algorithm does not need any coefficient, such as the discharge coefficient. Results obtained by testing the CSI algorithm on several examples were not impressive. They suggested that the specific nodes representing the control structure should be modified to simulate the flow in the canal accurately, because the control structures coincide with the bends of the canal (Figure 7.5) and a loss coefficient should be introduced to simulate the energy losses in the canal bend. For this reason, a new kind of node was built for this test, and the equations involved in this specific node are defined as:

$$\left.\begin{aligned} & A(y_{R1})\times v_{R1} - A(y_{R2})\times v_{R2} - q(K) = 0 \\ & y_{R1} + \frac{v_{R1}^2}{2g} - y_{R2} - \frac{v_{R2}^2}{2g} - Loss1 - Loss2 = 0 \end{aligned}\right\} \quad [7.1]$$

$$Loss1 = c_T \times \frac{\left|v_{R1}^2 - v_{R2}^2\right|}{2g}$$

$$Loss2 = L_T \times \frac{\left[S_f\left(b_1, y_{R1}, v_{R1}\right) + S_f\left(b_2, y_{R2}, v_{R2}\right)\right]}{2} \qquad [7.2]$$

$$S_f\left(b, y, v\right) = n^2 \frac{v^2}{\left[\frac{by}{b+2y}\right]^{\frac{4}{3}}}$$

where y_{R1}, v_{R1}, y_{R2} y v_{R2} are the water level and velocity upstream and downstream of the bend, respectively, and the transition length is 4 m. The bottom widths are b_1 and b_2 (upstream and downstream of the bend, respectively; in our case, $b_1 = b_2$). The coefficient of energy lost in transitions is c_T, n is the Manning coefficient, S_f is the Manning equation and q(K) is the water withdrawals or extracted flow during the period K.

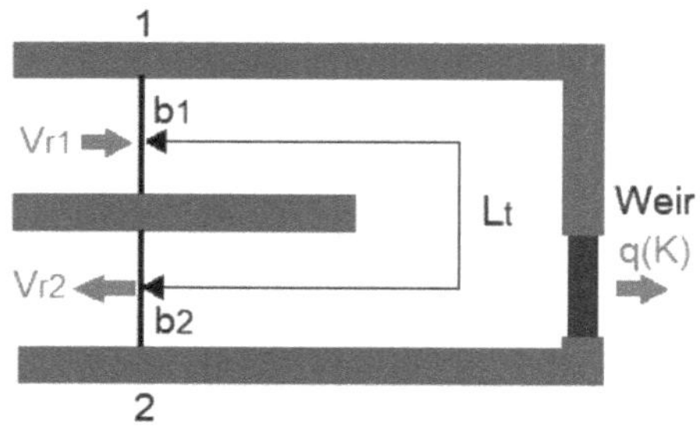

Figure 7.5. *Schematic diagram of the top view of a canal bend of the Canal PAC-UPC*

The extracted flow q(K) is calculated by the CSI algorithm in each regulation period, and every term is compiled in the extracted flow vector.

We calibrated the physical parameter (c_T) by conducting several tests. We simulated the water profile of the canal with a computer model from the initial conditions introduced in section 7.4; therefore, we conducted these tests in the real canal under the same flow conditions and took water-level measurements in several sections of the canal. The physical parameters of the model were adjusted until the same water levels in these sections of the canal were obtained in both cases (simulated and real).

The physical parameters adjusted in each canal bend are shown in Table 7.2:

Specific loss node	Upstream pool	Downstream pool	c_T
1 (from L2 to L4)	1	2	0.2
2 (from L4 to L6)	2	3	0.2
3 (from L6 to L8)	3	4	0.2
4 (from L8 to L10)	4	5	0.2

Table 7.2. *Local loss coefficient for the canal bends.*

7.1.1. *Gates*

The canal has vertical sluice gates made of methacrylate reinforced with a metal skeleton in order to provide enough stiffness and to reduce weight. The vertical movement of the gates is guided by metal frameworks embedded in the canal and is executed by three-phase servomotors, which are located on top of the gates and commanded by control boxes situated next to the canal. This particular gate motorization only enables constant speed movements of about 1 cm/s. In this test, all gates except the first were kept out of the water during the whole test.

7.1.2. *Weirs*

Rectangular weirs are used to extract water in order to emulate the effect of offtake discharges in real irrigation canals. It is possible to obtain the flow rate extracted by measuring the water level upstream of the weir. Figure 7.6 shows a photo of a weir in the canal.

Figure 7.6. *Photo of a weir of the Canal PAC*

These weirs have a width of approximately 39 cm and were constructed starting from a 35 cm canal height, except for the end weir (weir 4), which starts from the canal invert. From this minimum height of 35 cm, it is possible to increase the height of the weir by placing PVC pieces in metal rails, one on top of the other. There are pieces of 5, 10, 20 and 35 cm. With different combinations, it is possible to cover a broad range of weir heights and, thereby, to achieve a broad range of output flows through the weir.

7.1.3. *Water-level sensors*

The canal currently has nine level sensors located in strategic places, upstream and downstream of each sluice gate and close to every rectangular weir (see Figures 7.1 and 7.2 for details). Their mission is twofold: sensors take measurements of water level to supply data to the CSI algorithm, and the measurements upstream of the weir are used to calculate the discharge through the weirs.

7.2. Initial and boundary conditions

We introduced the initial conditions from the water-level measurements at the checkpoints, the flow downstream of every pool (Table 7.5) and the position of gate 1. The upstream boundary condition is determined by gate 1 and the water level at the reservoir. The gate equation is represented for the next system of equations as:

$$
\begin{aligned}
A(y_s)v_s &= c_G \times a_G \times u \times \sqrt{L1(t) - y_s} \qquad \textit{(submerged hydraulic jump)} \\
A(y_s)v_s &= c_G \times a_G \times u \times \sqrt{L1(t) - Cc \times u} \qquad \textit{(unsubmerged hydraulic jump)}
\end{aligned}
\qquad [7.3]
$$

where y_S is the downstream water level of the reservoir; v_S is the velocity in that cross-section; $L_1(t)$ is the water-level measurement upstream of the reservoir; Cc is the contraction coefficient; C_G is the discharge coefficient, which depends on different parameters, such as the upstream water level, gate opening, contraction coefficient and flow condition; a_G is the gate width and "u" is the gate opening.

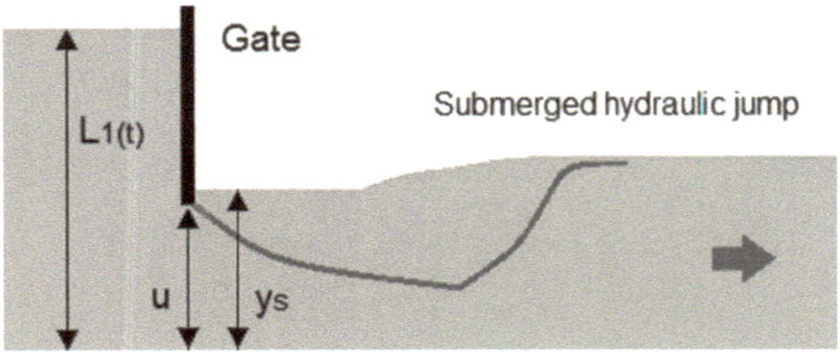

Figure 7.7. *Submerged hydraulic jump*

The gate features are shown in Table 7.3.

Gate	Numerical node upstream	Numerical node downstream	Gate discharge coefficient	Contraction coefficient	L_1 water level reservoir	Gate width (m)	Height of the gate opening (m)	Steep (m)
1	0	1	0.68	0.60	1.257	0.434	0.122	0.0

Table 7.3. *Features of the gate*

Finally, the downstream boundary condition is associated with weir 4, where the water level reaches a critical depth. The weir equation is represented by the following discharge equations of a sharp crested weir:

$$A(y_{R1}) \times v_{R1} - c_W \times B \times (H)^{\frac{3}{2}} = 0$$
$$c_W = \frac{2}{3} \times \sqrt{2g} \times c_d \qquad [7.4]$$

where y_{R1} and v_{R1} are the water level and velocity at the sharp crested weir; H is the measured head above the crest, excluding the velocity head; C_d is the discharge coefficient and B is the weir width.

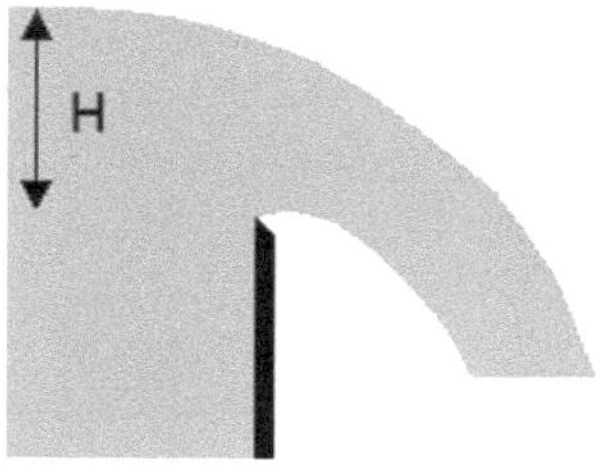

Figure 7.8. *Sharp crested weir*

The features of the sharp crested weir are shown in Table 7.4.

Weir	Numerical node	Weir discharge coefficient (C_W)	Weir height (y_W) (m)	Weir width (m)
4	341	0.5776	0.35	0.39

Table 7.4. *Features of the weir*

The steady state is the initial condition for the canal. The total flow is 110 l/s, and weirs 1, 2 and 3 are not operative. Table 7.5 shows the water-level measurements obtained manually and the flow rate at particular points at the initial time.

Checkpoint	Initial water level (m)	Flow (m^3/s)
1 (L4)	0.758	0.110
2 (L6)	0.730	0.110
3 (L8)	0.689	0.110
4 (L10)	0.644	0.110
5 (L11)	0.604	0.110

Table 7.5. *Initial conditions (water level at particular points)*

The initial steady profile for the canal is shown in Figure 7.9.

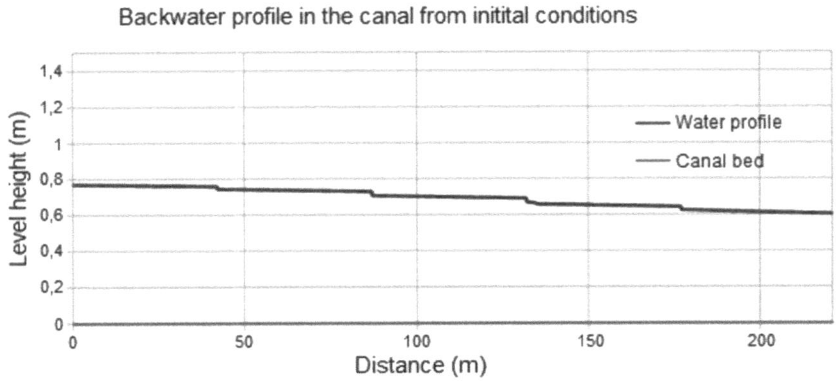

Figure 7.9. *Backwater profile for the canal from initial conditions (steady state). For a color version of the figure, please see www.iste.co.uk/bonet/canals.zip*

7.3. Discretization of a problem

The past time horizon is T^{KF} = 3,410 s = 57 min. The regulation period or the time interval between successive control actions for a controller is ΔT = 10 s, so the number of regulation periods during the predictive horizon is γ = 341. The canal has been spatially discretized (Δx) every meter, so the number of computational cross-sections along the canal is 220.

7.4. Scenario

The canal state has been perturbed by a disturbance using the weirs. The main objective of the CSI algorithm is to find the flow trajectory that better describes the flow through the weir.

After conducting several tests, we conclude that water-level sensors should be close to the crest weir, as the big changes in water levels are associated with these extraction points. The crest weirs are also close to the canal bends, and the local phenomenon introduced by the canal bends, which affects the backwater profile, should be taken into account. The local phenomenon has a significant influence on the value of the water-level measurements as it causes over-elevations in the backwater profile, which is a big problem for CSI. This is the reason why we only use sensors L6, L10 and L11 in CSI.

As described in Chapter 6, we use a one-dimensional computer model, whose flow is clearly two-dimensional in the canal bend. For this reason, we had to introduce a specific new node at the location of the canal bend in the 1D computer model in order to simulate the energy losses that occurred in this short reach.

The output data of the sensors must be as accurate as possible because the noise associated with the water-level measures can invalidate the test.

At the beginning of the test, the flow rate in the canal is 110 l/s. At a particular time (250 s after starting the test), some of the pieces that made up the lateral weir (W2) were removed, so the weir height changed to 55 cm. Later, at 1,950 s, the weir was closed again.

The canal did not have a flow meter and therefore the exact value of the flow through weir 2 was not measured directly. However, we could estimate the flow through weir 2 (see Figure 7.11), because we obtained the water-level measurements of the L7 sensor (Figure 7.10), which is the closest sensor to weir 2 (W2), and the discharge coefficient of weir 2 was calibrated in previous work (see Horváth [HOR 13]).

Table 7.6 shows the changes obtained with changing weir height in order to introduce the disturbance in the canal:

Time (s)	G1 Gate opening (m)	W1 weir height (m)	W2 weir height (m)	W3 weir height (m)	W4 weir height (m)
0	0.122	0.90	0.90	0.90	0.35
250	0.122	0.90	0.55	0.90	0.35
1950	0.122	0.90	0.90	0.90	0.35
3410	0.122	0.90	0.90	0.90	0.35

Table 7.6. *Disturbance characteristics introduced into the system*

Figure 7.10 shows the exact time when we introduce disturbance in the canal because the water level at sensor L7 decreases quickly by 3 cm.

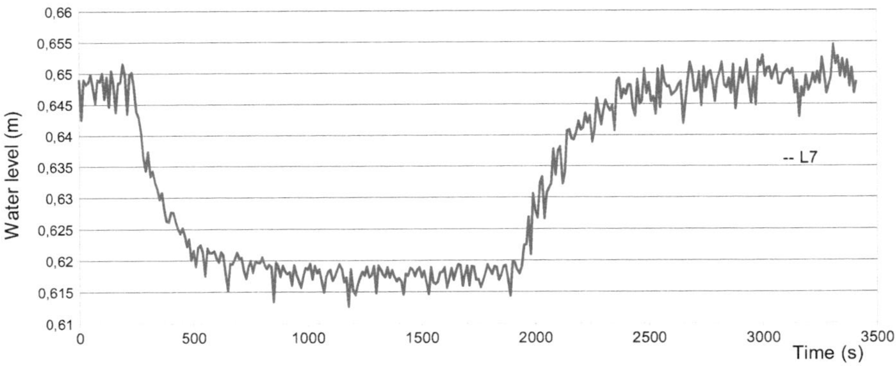

Figure 7.10. *Water-level measures at sensor L7*

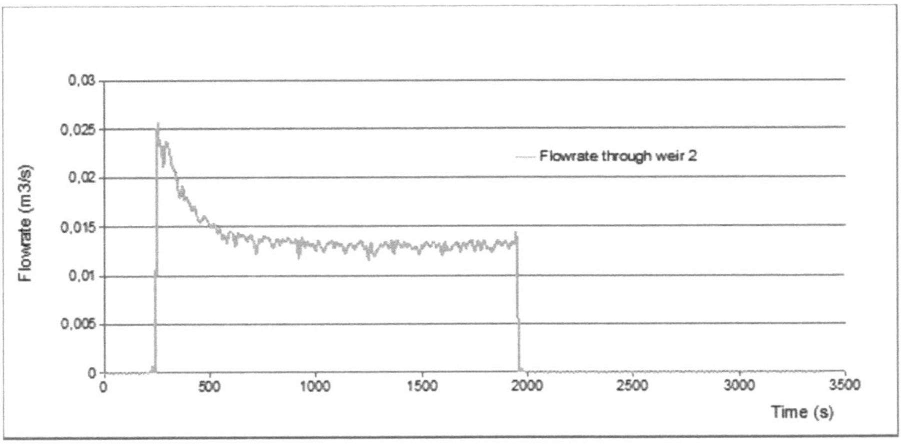

Figure 7.11. *Flow through weir 2 using the weir equation and the discharge coefficient calibrated by Horváth [HOR 13]*

Obviously, the flow extracted through the weir modified the water-level surface and the flow along the canal. From sensors L6, L10 and L11, we obtained the water-level measurements shown in Figure 7.12.

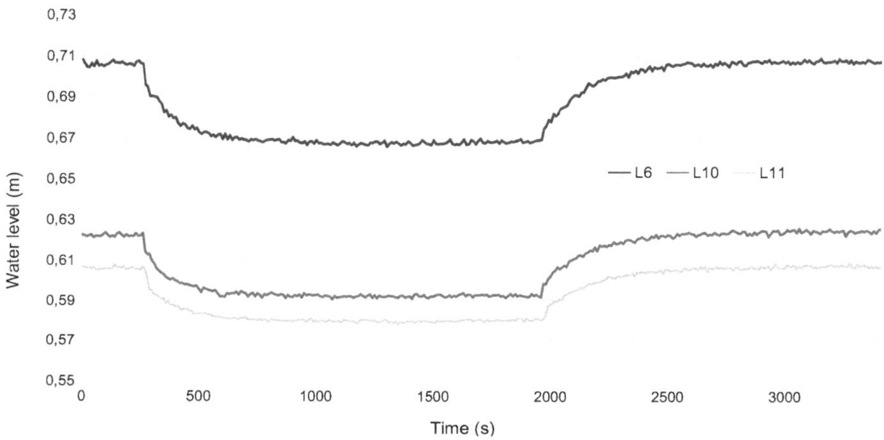

Figure 7.12. *Water-level measures at sensors L6, L10 and L11. The measures were obtained every second, and the average value was calculated every 10 s. For a color version of the figure, please see www.iste.co.uk/bonet/canals.zip*

The water-level measurements in sensors L6, L10 and L11 have a depth gage error of 1/2 mm, as can be seen in Figure 7.12, but the water-level measurements in sensors L3, L5 and L7 (Figure 7.10) have a depth gage error of 7/10 mm. These last measurements are not quite accurate for CSI in this test, because the water level variation in the canal caused by the flow through weir 2 is about 2.5–4 cm (Figure 7.12).

The water-level measures at the reservoir (L1) are shown in Figure 7.13. This data series, collected by sensor L1 every second, is used by the CSI algorithm to establish the upstream boundary condition. Although the water level at the reservoir is constant in time, due to the limitations of laboratory installations, the water level at the reservoir is variable in time, because it depends on the water level downstream of the gate. For this reason, the water level in the reservoir decreases when the flow is extracted through the weir, as shown in Figure 7.13.

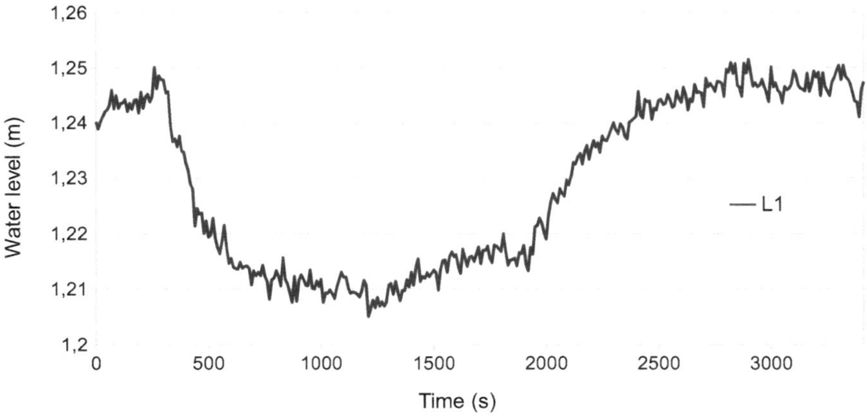

Figure 7.13. *Water-level measures at sensor L1 (upstream reservoir). The measures were obtained every second and the average value was calculated every 10 s*

7.5. Test results

The disturbances are introduced into the system by modifying the weir height. The sensors L6, L10 and L11 take the water-level measurements, which are introduced to the CSI algorithm to calculate a discharge through the weir that generates a variation in water levels at the checkpoints equal to the water level measured at sensors L6, L10 and L11. The extracted hydrograph explains the evolution of the water-level measures at the sensors during the past time horizon.

The hydrograph obtained by the CSI algorithm was filtered using a 10-period simple moving average (SMA) (see Figure 7.14). Figure 7.14 shows the flow through the weir (W2), obtained with the equation of a sharp-crested weir from the water-level measures at sensor L7 and the discharge coefficient calibrated by Horváth [HOR 13], which can be compared to CSI's hydrograph.

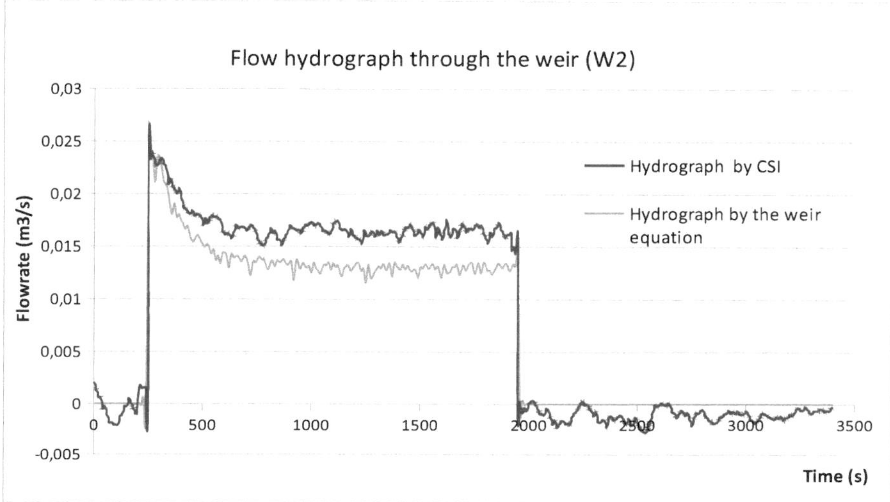

Figure 7.14. *Extracted flow rate through weir 2 by CSI using 10-period simple moving average versus extracted flow rate through weir 2 by the weir equation, using the discharge coefficient calibrated by Horváth [HOR 13]. For a color version of the figure, please see www.iste.co.uk/bonet/canals.zip*

After analyzing Figure 7.14, the following conclusions can be drawn:

– The algorithm obtains an extracted hydrograph similar to the real flow extracted through the weir (W2), especially at the initial moment, to introduce the disturbance.

– Although the real extracted flow was obtained by the weir equation using inaccurate water-level measures (sensor L7), these measurements are quite accurate for using this equation. As shown in Figure 7.14, the noise error of this curve is not significant.

– The water-level measurements of sensors L6, L10 and L11 were used as input data in computations of CSI because the variations of water-level measures are lower than with other sensors, as these sensors are far away from the canal bend.

– The flow rate difference between the hydrograph obtained by CSI and the hydrograph obtained using the discharge coefficient calibrated by Hórvarth was about 2.5 l/s.

– Probably, the differences between both hydrographs is the result of deviations in water-level measurements at sensors L6, L7, L10 and L11. However, it is also possible that these differences are due to a bad calibration of the equivalent Manning coefficient or due to the energy loss transition coefficient, although we tried to adjust them to be as efficient as possible. We have to point out that the differences between both hydrographs are not significant – only 2.5 l/s over 110 l/s along the canal.

– The initial conditions provide input data for the algorithm, but sometimes there is no possibility to start the simulating horizon knowing the initial conditions. For this reason, we test this example for different initial flow conditions, and the algorithm is sensitive to these initial conditions, which affects the flow rate extracted through the weir calculated by the CSI algorithm. However, if we choose a past time horizon of a reasonably long period, the influence of initial conditions on the canal flow disappears. The duration of this period depends on the geometry and the features of the canal.

7.6. Sensitivity of the extracted flow to the Manning coefficient

This test was performed in order to study the sensitivity of the flow through the weir to the Manning roughness coefficient of the canal. With this objective in mind, the same test was conducted three times, modifying the Manning coefficient more in each test by increasing the real value by 0.003 and 0.006 and decreasing the real value by 0.003. The solution obtained by the CSI algorithm is shown in Figure 7.15. This type of test, where the physical parameters of the canal are customized around the real value, was called the “tuning test” in Test Cases [CLE 98].

Figure 7.15 shows how the Manning roughness coefficient has a large effect on the solution obtained by the algorithm, so the Manning roughness coefficient value should be carefully selected before being introduced into the CSI algorithm.

The accuracy of the Manning coefficient is more important when the flow rate through the weir is low, because a little variation in its value can result in a large variation in flow rates similar to the flow extracted through the weir. There is a linear relationship between the increment of the Manning coefficient and the flow extracted through the weir.

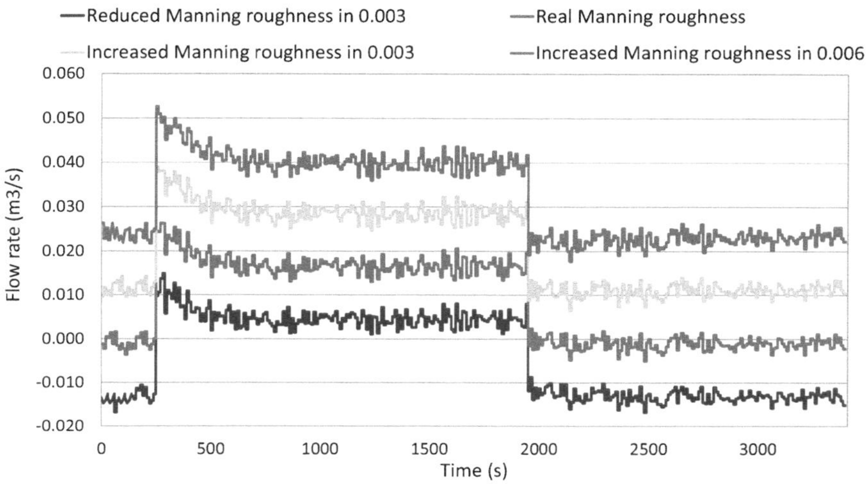

Figure 7.15. *Extracted flow rate at weir (W2) modifying the Manning roughness coefficients. For a color version of the figure, please see www.iste.co.uk/bonet/canals.zip*

7.7. Conclusion

Our conclusions are listed below:

– We have obtained a fast convergence and accurate results with the CSI algorithm, which has been possible thanks to the Levenberg–Marquardt method.

– The CSI algorithm is very stable because it calculates the extracted hydrograph very similarly to the real hydrograph extracted from the weir. That is, the result is physically feasible and the algorithm does not look for incoherent solutions, which could also define the canal state. For this reason, it is not necessary to introduce constraints to stabilize the algorithm.

– The water-level measurements must be as accurate as possible, as the result (extracted hydrograph) obtained by the CSI algorithm will be as accurate as the input data (water-level measurements).

– The CSI algorithm is highly sensitive to some physical parameters such as the Manning roughness coefficient and, by extension, other parameters, such as the local energy losses in the canal bend. For this reason, the algorithm should not be used in canals whose physical parameters are not well-known.

8

Online Predictive Control: GoRoSoBo

One of the main objectives in this book is to manage and drive the canal to the desired state during the irrigation periods. There are different types of algorithms able to manage a canal, for instance, feedforward algorithms, heuristic algorithms or feedback algorithms.

Feedforward algorithms are used in many canals but they are inconsistent in a system where the scheduled demand is not fulfilled by the users. Feedforward algorithms calculate the optimal gate trajectories taking into account a particular scheduled demand established for an irrigation cycle. The problem is when someone introduces a disturbance and the algorithm fails to manage the canal.

The heuristic algorithms are not based on physical laws, so the system remains a black box for them. These kinds of algorithms are usually used when large time series of measurements are available and the system is too complex to model. The problem with these algorithms is similar to the feedforward algorithms: when someone introduces a disturbance and the algorithm operates out of the range that was calibrated, the controller fails.

We have developed a feedback algorithm (GoRoSoBo; Gomez, Rodellar, Soler and Bonet) which gets the measured values and compares them with a reference, representing the desired performance. Any deviation from the reference is fed back into the control system to reduce the deviation of the controlled quantity from the reference. We introduce GoRoSoBo in this chapter.

8.1. Introduction

To fulfill our objective, we will introduce our predictive control which is part of the overall control diagram of an irrigation canal, introduced in Chapter 4. We introduce our developed algorithms in the diagram in Figure 8.1.

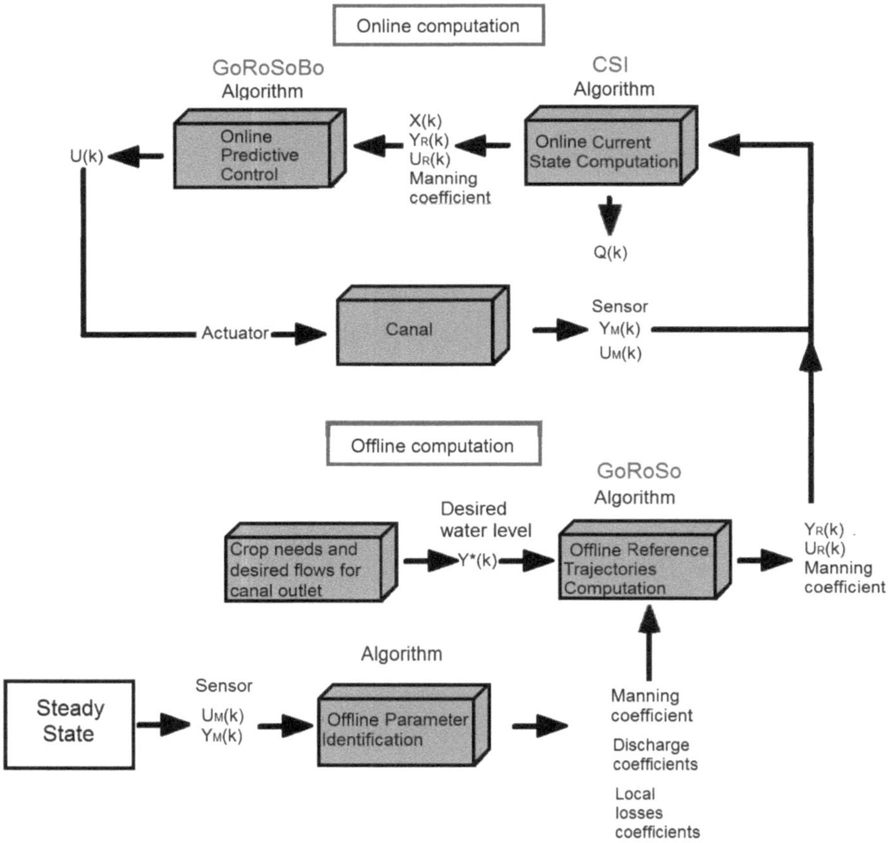

Figure 8.1. *Algorithms developed in this book in an overall control diagram of an irrigation canal. This cycle is done in each operation period, and GoRoSoBo calculates the optimum gate trajectories in each operation period for a predictive horizon*

The primary objective for a control algorithm is to maintain the water level at several points of the canal, where the flow is extracted by control structures (lateral weirs, pumps, offtake orifices). Typically, almost all of these structures are powered by gravity, so the flow control depends on the water level. Maintaining the water level at the checkpoints allows for control of the demand deliveries and protects the canal from significant variations in the water level which could damage the canal.

The objective of maintaining a constant water level at checkpoints is rarely achieved because the water deliveries along the canal are not always well implemented, and sometimes, these deliveries are unknown. The disturbances produce transients in the canal and, consequently, changes in water level and discharge at the orifice offtakes. The flow rate delivered differs from the scheduled delivery established by the watermaster, and as a consequence, the efficiency of the irrigation canal decreases. It is important to develop a feedback controller which returns the canal to the desired water level at the checkpoints as soon as possible.

Using feedback controllers is not synonymous with success. There are some feedback controllers which work in parallel with computer models, and the results obtained are good, but only in cases when the disturbance is small and only affects the immediate neighborhood of a section. As these controllers do not know the unscheduled deliveries introduced in the canal, if the disturbance is significant, the data introduced to the feedback controller that were previously obtained by the computer model are not good enough and the controller will fail.

These controllers also do not work correctly when the water level measurements at a checkpoint are far away from the desired water level. The gate trajectories are far away from the real solution because the disturbance has modified the canal state entirely and the controller only takes into account some measured values at several checkpoints in the canal. These controllers only consider the perturbation indirectly, through their effects on the water level measurements of the system.

GoRoSoBo does not have this problem because we have developed the CSI algorithm to estimate these perturbations and establish the hydrodynamic canal state at every cross-section, as we have seen in Chapters 5–7. All these data can be sent to the feedback controller (GoRoSoBo) to calculate the optimum gate trajectories, taking into account a particular target which is usually to maintain the water level at the checkpoints.

There are other authors, such as Delgoda *et al.* [DEL 13] and Van Overloop *et al.* [VAN 08], who have also developed predictive controllers based on predictions of unscheduled flows.

In a particular case or "**ideal case**", where there are no disturbances in the canal, a feedforward algorithm could manage the canal, as we show in Figure 8.2. In this state, the gate positions are fixed by a feedforward algorithm taking into account the scheduled demand (a particular desired water level at the orifice offtake), which is 4 m^3/s during the irrigation cycle, as shown in Figure 8.2.

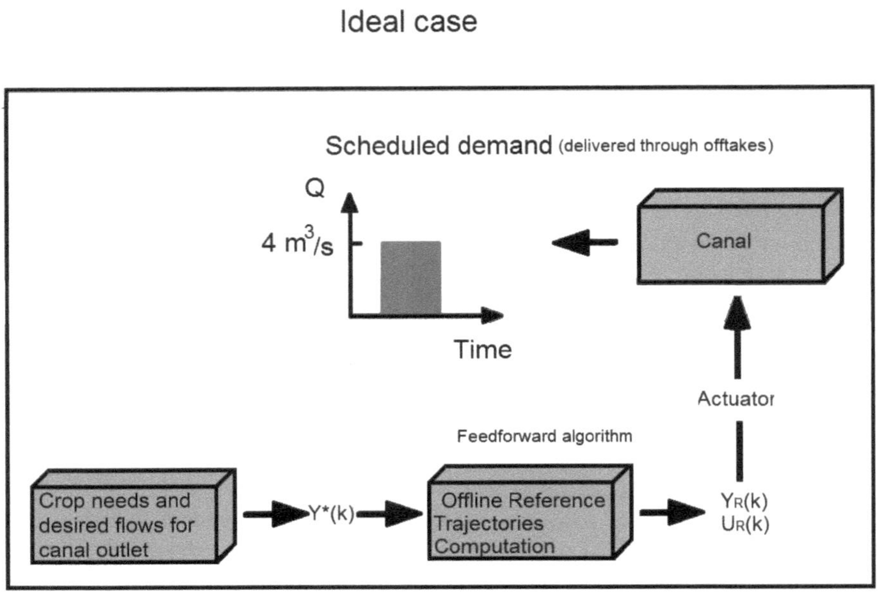

Figure 8.2. *Ideal case for a predictive control which maintains a water level to satisfy a scheduled demand of 4 m^3/s*

So the ideal case is often quite far from reality because someone could pump an unknown flow such that the water level in the offtakes orifice decreases, as does the flow delivered to the farmers, see Figure 8.3. In this example, someone introduces a disturbance (2 m^3/s) in the canal and the water level at the checkpoints decreases. The feedforward algorithm only considers the scheduled demand previously fixed, so the water level at the checkpoints is not recovered and the flow delivered to the farmers is reduced as well.

In this particular case, our overall control diagram of an irrigation canal is mandatory, as we have introduced before (Figure 8.1). Initially, the canal shows a steady state (Figure 8.4(a)), but someone introduces a disturbance at T=40 s (Figure 8.4(b)). The water level measurements are sent first to the CSI which calculates the disturbances and the current canal state, and then GoRoSoBo calculates a new gate trajectory to fulfill the objective (a return to the desired water level at the offtake) (Figure 8.4(b)). The new gate trajectory is sent to the actuator which modifies the gate position (Figure 8.4(c)). The canal state then returns to the desired state (Figure 8.4(d)).

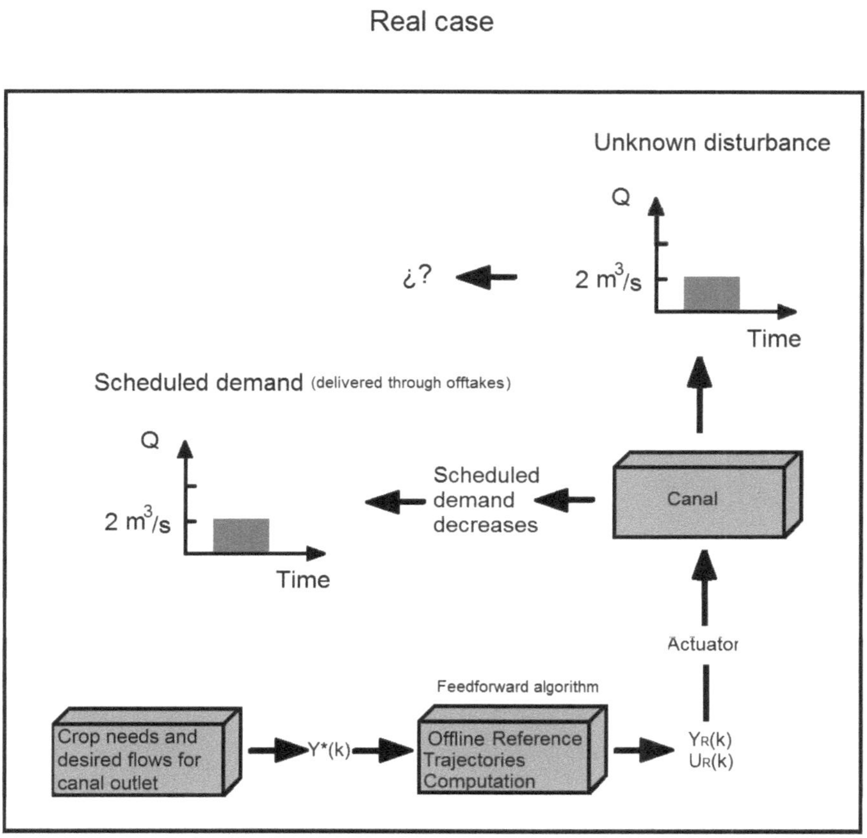

Figure 8.3. *Real case for a predictive control. The controller calculates the gate trajectories to maintain the water level and supply the scheduled demand of 4 m^3/s. The problem is when someone introduces a disturbance of 2 m^3/s. The discharge through offtakes would be reduced by 2 m^3/s*

The question is why have we developed a new feedback algorithm when there are already a lot of them in the literature? Because while some of them, such as CLIS [LIU 98] or PILOTE [MAL 95], operate quite well and obtain good performance indicators in the Test Cases proposed by ASCE [CLE 98], they do not operate precisely as we want, and other algorithms do not achieve our objectives, as we describe below:

– Many feedback algorithms calculate the optimal gate trajectories from empirical relations between the gate trajectory and the water level error at checkpoints obtained in specific canals. But these algorithms are not based on physical laws, and the empirical relations only operate in a particular range.

– There are other feedback algorithms which calculate the optimal gate trajectories from relations between the gate trajectories and the water level error in a particular steady state. In cases when a small disturbance is introduced in the canal, these algorithms operate quite well. In cases when the disturbance is significant, the initial steady state is quite different to the current canal state, so the gate trajectories calculated with these algorithms will not be the optimal trajectories.

– There are feedback control algorithms which need two or more water levels and flow measurements for each pool to calculate the gate trajectories, which is not always possible.

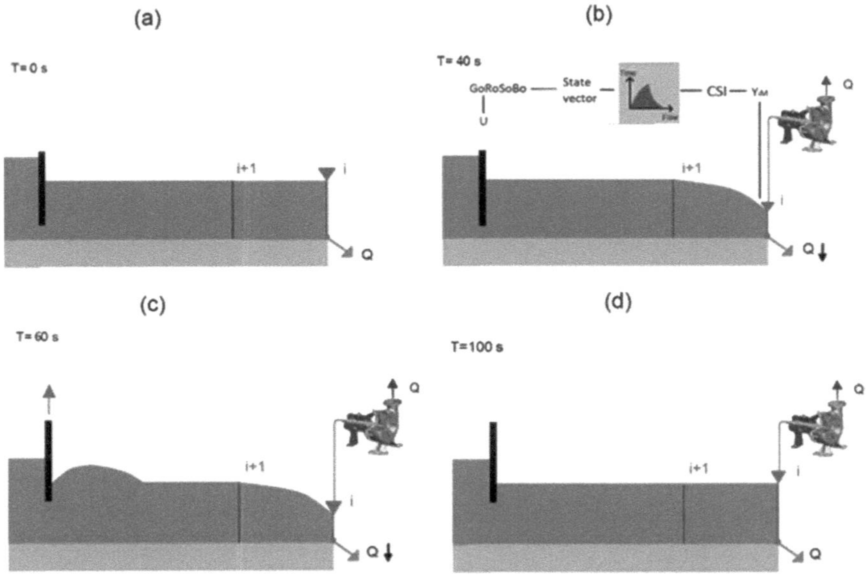

Figure 8.4. *Logic of control in our overall control diagram of irrigation canal*

In this book, our feedback algorithm (GoRoSoBo) calculates the optimum gate trajectories from a water level error with the aid of the HIM (Hydraulic Influence Matrix, Soler [SOL 03]). This matrix establishes the influence of a gate movement on the free surface. Every element of the HIM establishes the variation of the variable X (water level and velocity along the canal) associated with any change in gate positions [8.1]. The HIM can be recalculated at every time step, and thus we know the influence of a gate position on the water level at the current canal state:

$$[HIM(U)] = \left[\frac{\partial X}{\partial U}\right]. \quad [8.1]$$

The name of the Hydraulic Influence Matrix (HIM) originated due to the hydraulic components which represent the influence of gate trajectories over the state vector (see Chapter 5) at all sections during the predictive horizon. Based on the full Saint-Venant equations, and using the first derivative $\left(\frac{\partial X}{\partial U}\right)$ in an analytical process, the changes in flow behavior from a sluice gate movement at any section in a certain time step can be established.

To show a general view of the HIM, see Figure 8.5. When we open a gate in a canal, we modify the canal state, the water level and the velocity at particular points at a particular time. For instance, at T=10s, we modify the water level and velocity at cross-sections close to the gate (cross-section i), but we do not modify the hydrodynamic variables of a distant cross-section ("i+2" or "i+3"). Instead, at T=30s, the hydrodynamic variables in all cross-sections from "i" to "i+3" have been modified. The HIM establishes how a gate movement may modify the hydrodynamic variables in all sections of the canal during a prediction horizon.

One of the main numerical problems of the HIM is that it is an ill-conditioned matrix, as we introduced in Chapter 5, partly explained by the disparity of influence between its element values, as some of them are zero. The influence of a sluice gate movement on a particular section depends on the distance between them, and the influence decreases after several regulation periods from the gate movement.

In that sense, if we look at Figure 8.6, the gate at "i" changes its position (ΔU_i) at T=0, so the gate generates a perturbation which travels upstream and downstream of the canal. In this case, the section "i+3" at T=0s is not influenced by the gate movement (ΔU_i), but at T=30 s, it is influenced by the gate. The magnitude of the influence of ΔU_i is not equal in section "i" at T=0s and in section "i+3" at T=30s, because the influence of the perturbation decreases in space and time.

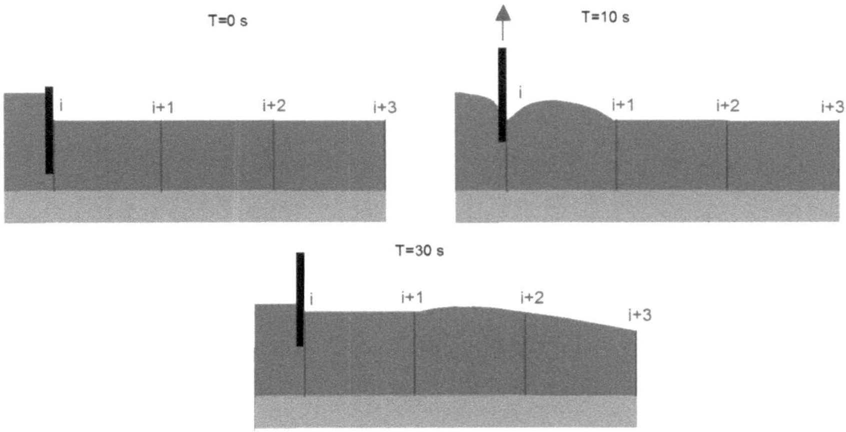

Figure 8.5. *A general view of the influence of a gate movement in the hydrodynamic state of a canal*

We can choose the space/time domain of the canal (Figure 8.5), as many sets of points in to form a computational grid where we can establish the influence of a gate movement on the water level and velocity at each of these points (Figure 8.6).

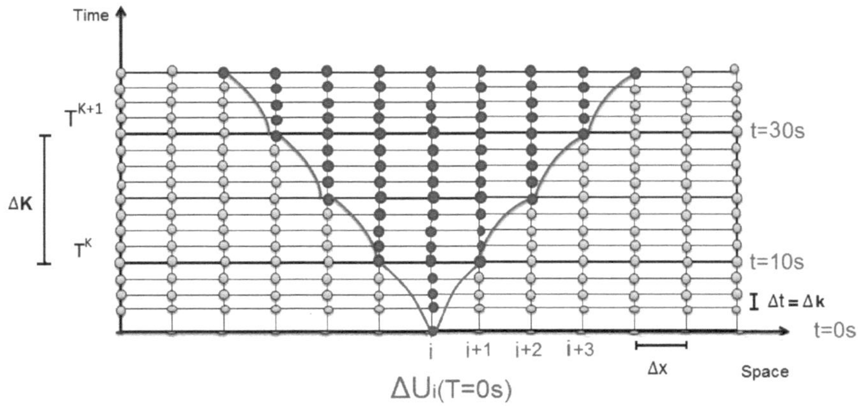

Figure 8.6. *Influence of a gate movement at several points of the space/time domain in the canal disposed in a computational grid. Notice that "K" with a capital letter denotes the time interval of control and "k" with a small letter denotes the time instant of simulation. For a color version of the figure, please see www.iste.co.uk/bonet/canals.zip*

8.2. Problem statement

We could calculate the perturbations associated with a particular gate movement in a canal from a computer model based on the Saint-Venant equations. The GoRoSoBo does the inverse process, that is, it calculates the gate trajectories that most closely approximate a particular perturbation (for example, water level error at checkpoint).

The GoRoSoBo algorithm is a feedback controller where the desired output (variable with a desired set point) is used to adjust the simulated water level to drive the system to the desired water demand from control actions (the optimal gate trajectories).

There are multiple ways to obtain the optimal gate trajectories, for instance, there are controllers that are based on the inversion of the equations describing the movement of the water; other controllers are based on the optimization of a given criterion for the present and a given time horizon and others are based on the use of the proportional/derivative/integral of the measured error to correct the control action, etc.

Our proposal defines an objective function from the total deviation of the data prescribed by the user (target water level) from the simulated data (the prediction output vector). This objective function is minimized through an optimization procedure with the objective to obtain the optimal gate trajectories. In that sense, the GoRoSoBo solves an inverse sequential problem as an optimization procedure.

8.2.1. *Definition of the optimization problem*

In this optimization problem, we replace the water level measurements with the prediction output vector ($Y_1^{k_F}$) obtained from a computer model (more details in Chapter 5) due to the water level measurements used previously by the CSI to compute the real extracted flow and the hydrodynamic canal state. The prediction output vector contains the values of the water level at certain cross-sections of the canal for a prediction horizon ($1 - k_F$) according to particular gate trajectories. The dimension of the prediction output vector is $n_Y = k_F \times n_c$, where k_F is the final time step of the prediction horizon and n_c is the number of checkpoints.

The GoRoSoBo calculates the optimal gate trajectories during a prediction horizon. As illustrated in Figure 8.7, the gates are operated with a period ΔK. Gate trajectories are assumed as piecewise functions. The gate position remains fixed during the regulation period and the gate position could change between regulation periods. The gate trajectory vector is defined by combining the gate positions as follows:

$$U = \left[U_1(1), \ldots, U_{n_g}(1), \ldots, \ldots, U_1(K_F), \ldots, U_{n_g}(K_F) \right]^T, \quad [8.2]$$

where the dimension of this vector is $n_U = K_F \times n_g$, n_g is the number of gates and K_F is the final regulation period. In each operation period, the GoRoSoBo calculates new gate trajectories for a prediction horizon.

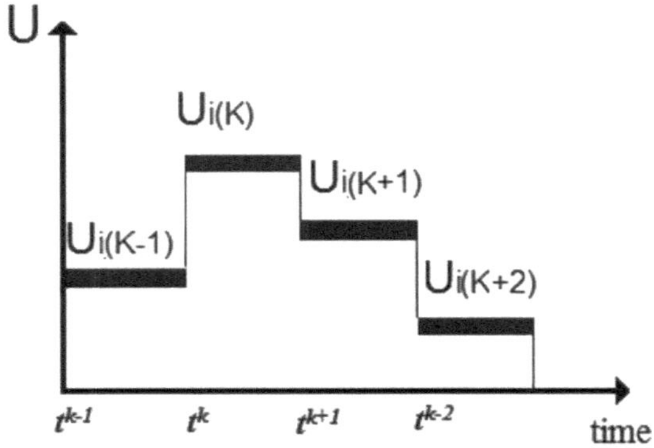

Figure 8.7. *Mathematical representation of a gate trajectory*

The data prescribed by the user in this optimization problem are the water level at checkpoints, the location of gravity offtakes and the algorithm that has to maintain the water level.

We can write the target water level vector at a particular time step k as

$$y^*(k) = \left[y_1^*(k), \ldots, y_i^*(k), \ldots, y_{n_C}^*(k) \right]^T. \quad [8.3]$$

Finally, all vectors [8.3] for each time instant of the predictive horizon are joined to define the desired water level vector, the dimension of which is n_Y. We define this vector as

$$Y^* = \left[y^*(1), y^*(2), \ldots, y^*(k_F - 1), y^*(k_F) \right]^T. \quad [8.4]$$

We choose a set of points in the space/time domain to form a computational grid where the solution is computed using a numerical procedure (Figure 8.8). We can check the desired water level vector values in the computational grid in Figure 8.8 (dots) as well as the measured water level vector (circles).

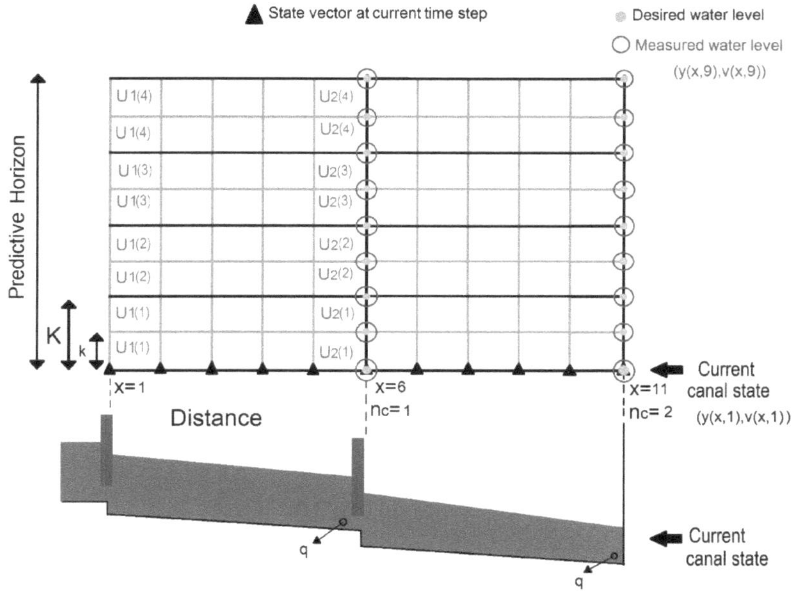

Figure 8.8. *Numerical grid in a canal with two pools controlled by two checkpoints downstream of each pool. There are two sluice gates. Gate trajectories are defined with four operation periods. The x/t-dots where the flow behavior is defined are shown. For a color version of the figure, please see www.iste.co.uk/bonet/canals.zip*

8.2.2. *Control problem: objective function*

As discussed in Chapter 3, an objective function expresses the aim of the control model to minimize or maximize variables to obtain an optimal response. An objective function can be the result of an attempt to express a control action (gate movement) in mathematical terms in order to drive the system to the desired output (variable with a desired set point).

In this way, we thus define ideal flow behavior as a sequence of desired water level values for given sections (checkpoints) at several time steps. In order to continue with the mathematical treatment, these values are defined by the desired water level vector.

In reality, the sequence of simulated water levels could be distant from the desired water level vector due to disturbances, and this variation between both values is defined by the objective function.

The objective function expresses the residual value between the prediction output vector and the desired water level at the target points. Minimizing this function, we obtain a set of control actions (U^*) and the best possible approximation to the desired value (Y^*). This is the goal of the objective function and U^* is the solution to

$$J(U) = \frac{1}{2}(Y_1^{K_F}(U) - Y^*)^T[Q'](Y_1^{K_F}(U) - Y^*) \qquad [8.5]$$

$$\text{minimum}\, J(U^*) = F\left[Y_1^{K_F}(U^*)\right], \qquad [8.6]$$

where Q' matrix is a weighing matrix with dimension $n_y \times n_y$. This matrix could be used to define the level of importance of the water level error in a particular checkpoint, weighting some values more than others, depending on their importance. We define this matrix as the identity matrix in the GoRoSoBo algorithm. The vector U contains the gate trajectories, and the dimension of this vector is $n_U = K_F \times n_g$, where $Y_1^{K_F}(U)$ is the water level at the checkpoints from K_1 to K_F due to gate trajectories U.

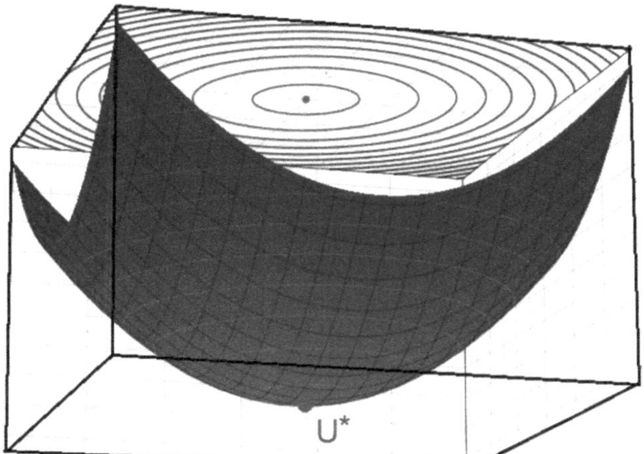

Figure 8.9. *Minimum value (U*) for an objective function. For a color version of the figure, please see www.iste.co.uk/bonet/canals.zip*

8.2.3. *Predictive control*

Once we know the water level/velocity along the canal at time step *K* with the aid of CSI (Figure 8.10), we can predict the water level/velocity along the canal for a predictive horizon (from *K* to *K*+3) as we know the canal state and the real extracted flow at *K*, the gate trajectories and the scheduled demand from *K* to *K*+3. In that sense, we can obtain the water level error at the checkpoints, the difference between the target value and the predicted water level, so that the predictive control could calculate some actions at *K* (present time) for the prediction horizon to remove the water level error. At the end of this process, we obtain a predicted output of the system at *K* for the prediction horizon, see Martín-Sánchez and Rodellar [MAR 96].

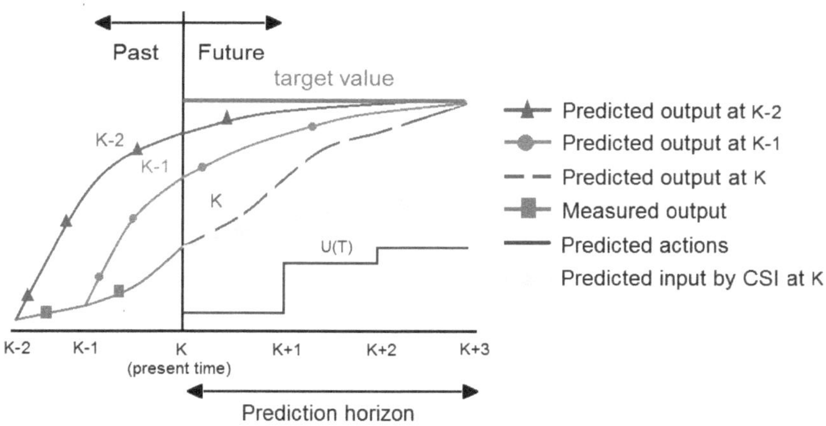

Figure 8.10. *A model predictive control scheme. For a color version of the figure, please see www.iste.co.uk/bonet/canals.zip*

In that way, we could establish the perturbation of the gate trajectory vector (ΔU) [8.7] to correct the water level error vector at the checkpoints (ΔY) for a prediction horizon from a simplification of the Hydraulic Influence Matrix (HIM(U)′) which represents the influence of the gates on the **water level vector** at checkpoints

$$\Delta Y = [HIM(U)']\Delta U$$

$$[HIM(U)'] = \left[\frac{\partial Y}{\partial U}\right]. \quad [8.7]$$

The HIM makes it possible to implement the GoRoSoBo algorithm, as this matrix establishes a direct relation between the water level error and the new gate trajectory vector. In that sense, the relation between the gate movements (ΔU) needed to restore the volume of water (striped area) in the canal shown in Figure 8.11 is established by the HIM.

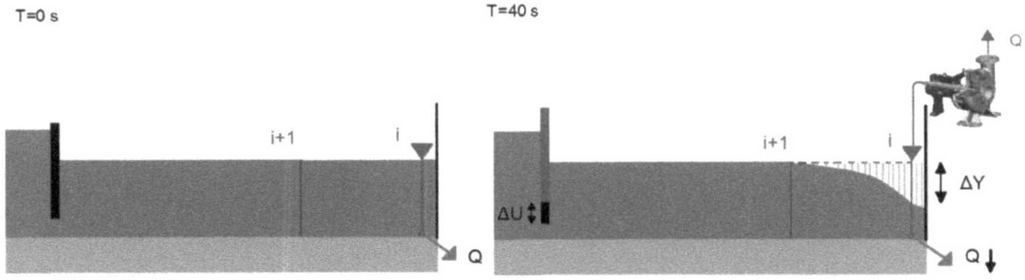

Figure 8.11. *Logic of control of the feedback algorithm*

The analytic process to obtain $\left(\frac{\partial Y}{\partial U}\right)$ analyzes the influence of a perturbation on the free surface described by the Saint-Venant equations in their characteristic form. If we look at Figure 8.6, a perturbation (ΔU) introduced in the space/time domain generates an upstream wave and a downstream wave which are both defined by characteristic curves; we could analyze the variation in the water level from the characteristic curves generated by the perturbation in an analytical process. Instead, the numerical process to obtain this matrix consists of the discretization of the analytic term, calculating the state vector X (water level and velocity along the canal) and considering a perturbation of the gate trajectories (U+ΔU) and the state vector X obtained from the gate trajectories (U) [8.8]. The analytic and numerical processes were analyzed by Soler [SOL 03], and both results were similar:

$$\frac{\partial Y_1^{k_F}}{\partial U(K)} \text{(analytic process)} \approx \frac{Y_1^{k_F}(U(K)+\Delta U)-Y_1^{k_F}(U(K))}{\|\Delta U\|} \text{(numerical process)} \qquad [8.8]$$

8.2.4. *Input/output variables of a controlled system*

The system input variables are applied to a control method to produce a control action and modify the output variables of the system. In our problem, the system input variables are the state vector at the current time step, scheduled demands, the desired water level vector and the gate trajectories in a previous predictive horizon. The system output variables would be the gate trajectories at the current time step for a prediction horizon with the objective of keeping the desired water level at the checkpoints. In Figure 8.12, we show the input and output variables in the GoRoSoBo algorithm, taking into account the considered variables in the process, as follows:

– Control method: GoRoSoBo algorithm.

– Input variables: scheduled demands, gate trajectories (previous prediction horizon), state vector at the current time step and the desired water level at checkpoints.

– Output variables: gate trajectories at the current time step.

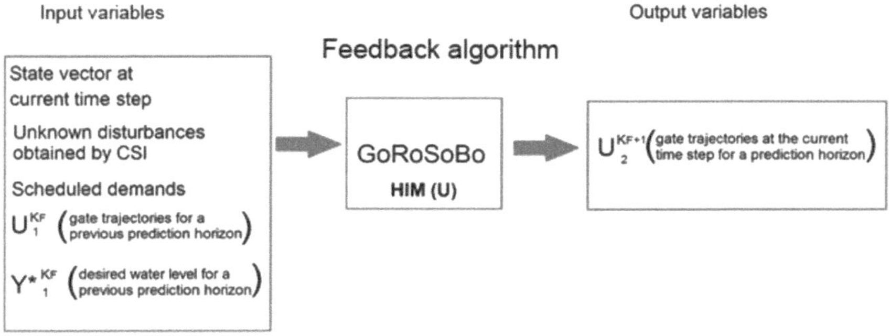

Figure 8.12. *Representation of the feedback algorithm*

8.2.5. *Analysis of the Hydraulic Influence Matrix: HIM(U)*

The HIM defines the influence of the gate on the canal state. The compiled HIM ($[\mathrm{I_M}(u)]_i$) for a gate (*i*) for a particular predictive horizon (*j*) is shown in the next equation:

$$[\mathrm{I_M}(U)]_{ij} = \begin{bmatrix} \dfrac{\partial X_1^k}{\partial U_i(1)} & \cdots & \dfrac{\partial X_1^{k_F}}{\partial U_i(K)} & \cdots & \dfrac{\partial X_1^{k_F}}{\partial U_i(\varphi)} \end{bmatrix}_j$$

$$= \begin{bmatrix} \dfrac{\partial x^1}{\partial U_i(1)} & \cdots & \dfrac{\partial x^1}{\partial U_i(K)} & \cdots & \dfrac{\partial x^1}{\partial U_i(\varphi)} \\ \vdots & \ddots & \vdots & \ddots & \vdots \\ \dfrac{\partial x^k}{\partial U_i(1)} & \cdots & \dfrac{\partial x^k}{\partial U_i(K)} & \cdots & \dfrac{\partial x^k}{\partial U_i(\varphi)} \\ \vdots & \ddots & \vdots & \ddots & \vdots \\ \dfrac{\partial x^{k_F}}{\partial U_i(1)} & \cdots & \dfrac{\partial x^{k_F}}{\partial U_i(K)} & \cdots & \dfrac{\partial x^{k_F}}{\partial U_i(\varphi)} \end{bmatrix} \qquad [8.9]$$

where the state vector at time step k is defined, respectively, as

$$x^k = \left(y_1^k \;\; v_1^k \;\ldots\ldots y_i^k \;\; v_i^k \;\ldots\ldots y_{n_s}^k \;\; v_{n_s}^k\right) \qquad [8.10]$$

When there is more than one gate to control the canal, the matrix has to compile the gates all together. In that case, the HIM is compiled as

$$[\mathrm{I_M}(U)]_{Xj} = \left[\left[\frac{\partial X_1^{k_F}(U)}{\partial U_1(1)} \cdots \frac{\partial X_1^{k_F}(U)}{\partial U_{n_g}(1)}\right] \cdots \left[\frac{\partial X_1^{k_F}(U)}{\partial U_1(\varphi)} \cdots \frac{\partial X_1^{k_F}(U)}{\partial U_{n_g}(\varphi)}\right]\right] \qquad [8.11]$$

where sub-index X is used to denote the HIM on the state vector.

The dimension of $[\mathrm{I_M}(U)]_{Xj} = n_{HX} \times n_{HY}$, $n_{HX} = 2 \times n_s \times (k_F - k_I)$, $n_{HY} = \varphi \times n_g$, n_s is the total number of cross-sections that have been discretized in the canal, φ is the number of regulation times of the prediction horizon, n_g is the number of gates, i is the "ith" gate, j is the "jth" prediction horizon and the dimension of the state vector (x) at the time step k is $2 \times n_s$.

A feature of this matrix concerns the number of columns and rows. There are more rows than columns because every term $\frac{\partial x^k}{\partial U_i(K)}$ is a vector with $2 \times n_s$ rows, and the temporal discretization of the simulation Δt^k is smaller than the regulation period ΔT^K.

If the matrix only defines the influence of a gate on the water level (not considering the velocity), the number of elements of the vector is reduced to n_s.

Another feature of the HIM is that it shows some elements with a zero value, because there is no influence from the gates over the state vector at particular points during the prediction horizon which could destabilize the algorithm because the matrix is ill-conditioned. We obtain null values for some elements of the HIM because these elements express the influence of a disturbance at the present time when the disturbance is introduced in the future, or express the influence of a disturbance in a faraway section of the canal where the disturbance has no influence.

One of the main problems with the HIM is the time required to calculate the matrix. This depends on the PC features and the total parameters: gates, checkpoints and the predictive horizon. We tried to reduce the calculation time when using the HIM around a steady state obtained previously because the GoRoSoBo operates in real time and there is not much time to calculate the gate trajectories between time steps. This option was discarded at the beginning when we tried to solve some hard tests with the GoRoSoBo, such as the Test Cases proposed by the ASCE committee [CLE 98], and it was impossible to solve the problems and get reasonable performance indicators using the HIM in a steady state. The unscheduled water changes are quite huge in these tests, and the free surface after several time steps is quite different to the free surface at the initial steady state. In that sense, the HIM would not represent the influence of the gate trajectories on the current canal state.

The HIM around a steady state would be acceptable for a test without huge flow changes, although the time duration of the flow changes is also important. If the external disturbance is smooth, but its time duration is long, the gate trajectory obtained using the HIM around a steady state gets worst in each regulation period. For that reason, it would be impossible to control the canal.

We could compare the problems associated with the HIM around a steady state with a loop problem: as the steady state is not the current canal state, the HIM around a steady state does not reflect the real canal state, and the gate trajectories obtained with this matrix will not be accurate. As the gate trajectories are not enough accurate, the real water level will never reach the desired water level at target points.

8.3. Algorithm controller for unconstrained optimization

Initially, the GoRoSoBo solved an unconstrained optimization problem by minimizing an objective function to reach the solution. There are several methods to solve an unconstrained optimization problem. All of them have to reach the solution (U^*) and verify two conditions to be sure that U^* is the better solution. The first condition demonstrates that the solution is a maximum, minimum or a saddle and the second condition demonstrates that the solution (U^*) is a minimum. These conditions have to be achieved by all optimized methods (**constrained** and **unconstrained**). In this way, it is interesting to introduce these order conditions.

8.3.1. *First-order and second-order conditions*

An extensive literature exists about the order conditions in optimization problems, for instance, Fletcher [FLE 87] and Luenberger [LUE 84]. Before solving an optimization problem, we have to be sure that the optimization problem fulfills the first- and second-order conditions.

The first-order conditions are necessary to find an optimal solution for an unconstrained optimization problem such as

$$\nabla J_U(U^*) = 0 \qquad [8.12]$$

$$\nabla_U J(U^*) = (\nabla_U X_1^{K_F}(U^*)^T - Y^*)^T [C]^T [Q]\left([C] X_1^{K_F}(U^*) - Y^*\right) = 0 \qquad [8.13]$$

where $\nabla J_U(U^*)$ is the first derivative of the objective function around the solution (U^*).

In fact, many minimization methods are based only upon trying to locate a solution U^*.This may not be a strict local minimum and in general is referred as a stationary point. A second condition is necessary to confirm that U^* is a strict local minimum point. As we introduced previously, a minimum corresponds to a positive definite Hessian matrix.

Where the Hessian matrix can be defined as

$$\nabla^2_{UU} J(U^*) \approx (\nabla_U X_1^{K_F}(U^*)^T)[C]^T[Q][C](\nabla_U X_1^{K_F}(U^*)^T)^T \quad [8.14]$$

When $\nabla^2_{UU} J(U^*)$ is used in a simplified form, this equation is called the Gauss-Newton method or least squares method [8.14].

8.3.2. *Optimized problem theory*

As noted, the GoRoSoBo was initially developed to solve an unconstrained optimization problem using the Levenberg-Marquardt method. The theory associated with this method was explained in Chapter 5, when the CSI algorithm was introduced, although we have developed this numerical method in terms of Brogan perturbation theory (see Appendix 1) in which we use a Taylor series to develop the perturbation equation around a gate trajectory.

8.3.3. *Implemented algorithm*

We used the Levenberg-Marquardt method because this method is robust and applicable to ill-conditioned matrices such as the Hessian matrix. The HIM is a positive definite matrix but it is also an ill-conditioned matrix. The approximation of the Hessian matrix, obtained by multiplying the HIM by its transposition, has a worse condition number than the HIM. As noted, we use the Levenberg-Marquardt method which introduces a value (ν) in the diagonal of the Hessian matrix to improve the condition number of the matrix

$$H = [\nabla^2_{UU} J(U^*)] + \nu\,[I]) \quad [8.15]$$

where $[I]$ is the identity matrix and $\nu > 0$ is a real value named the Marquardt coefficient. The steps defined in the implemented algorithm are shown in the next flowchart which is a type of diagram that represents every process, showing the steps as boxes and their order by connecting them with arrows. This representation illustrates a solution to the optimization problem. Process operations are represented in these boxes and the sequencing of operations by the arrows. Before showing the flowchart, we have to define some matrices and variables that are used in the flowchart:

– [Q'] is the weight matrix.

– [C] is the discrete observer matrix (a matrix of zeros and ones).

– $r' = [C]X_{ki}^{kF}(U_i) - Y^*$ is the residual vector.

– $X_{ki}^{kF}(U_i)$ is the state vector.

– $[I_M] = I_M\left[X_{ki}^{kF}(U_i)\right]$ is the Hydraulic Influence Matrix, all evaluated in U_i.

All the processes of the algorithm are defined in each loop, where the solution U* is calculated in each regulation period (i) for a predictive horizon.

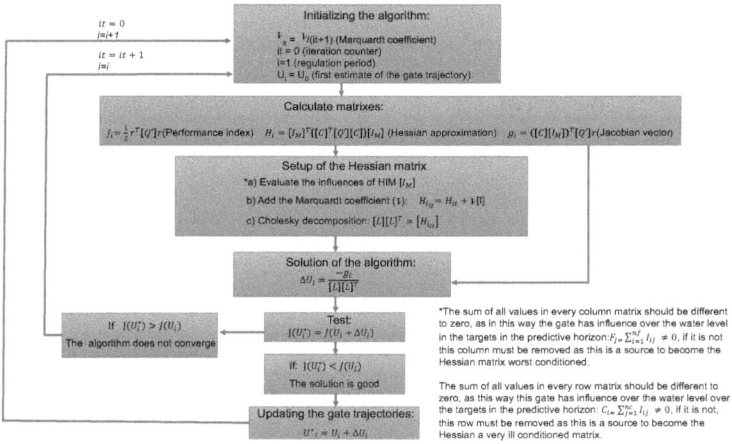

Figure 8.13. *GoRoSoBo scheduled: initial version for the unconstrained optimization problem algorithm*

8.3.4. *Analysis of the unconstrained optimization problem*

As introduced in section 8.3.1, our control algorithm was implemented from the HIM around a steady state solving an unconstrained optimization problem, but this method did not function properly in several tests in which the disturbances were significant.

This part of the book is important because we realized that it is impossible to develop a control algorithm without information about the current canal state. A water level error (difference between the computed water level and the measured water level) at a checkpoint in an instant of time modifies the free surface of a canal for a short time, whereas a flow change at a checkpoint during a regulation period modifies the free surface for a long time (Figure 8.14). Both situations are quite different, and the control algorithms should take into account one case or another depending on the situation, since the gate trajectories will be different in both cases. For that reason, it is important to obtain the disturbances with the CSI so that we can know when the free surface is modified at any time.

If a disturbance is introduced into the canal, the algorithm cannot control the canal by measuring the water level change and making a prediction of the future error without information about the disturbance. We realized that in that moment the controller needed the CSI algorithm, as the control algorithm needs to know the current canal state and disturbances at every regulation time.

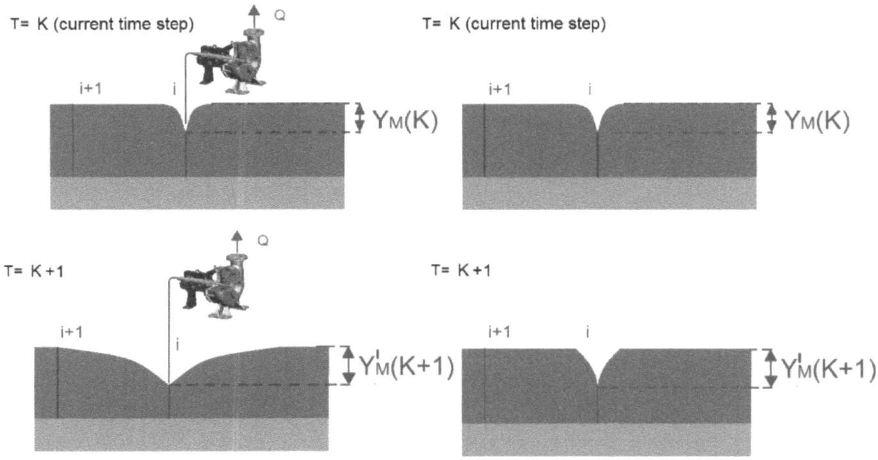

Figure 8.14. *Flow disturbance during a regulation period versus water level disturbance at a particular time step*

As previously introduced, our controller solves an unconstrained optimization problem based on the system of equations [8.7]. We implemented an unconstrained optimization problem in our predictive control because though it is easier to implement the code and reduce the computation time for the algorithms based on constrained optimization, the results with this algorithm were not as good as we thought.

Sometimes, one gets more information from failures than from successful tests. In cases in which significant unscheduled flow changes are introduced into the canal, the predicted water level error obtained by the CSI would be significant too. In such cases, the predictive controller would introduce important changes in gate trajectories to drive the system to the desired level such that changes would be physically impossible and the algorithm would fail. This was one of the reasons to introduce constraints in the optimization problem.

We realized the existence of this problem when we implemented the Test Cases proposed by the ASCE [CLE 98]. The first part of the Test Cases (scheduled offtake changes) has been solved by several feedforward controllers, one of which was GoRoSo [SOL 03]. Initially, this algorithm also solved an unconstrained optimization problem but GoRoSo was tested with these Test Cases and the results were not good due to the important scheduled offtake changes. This was one of the reasons for modifying the GoRoSo algorithm and implementing a constrained optimization problem which would successfully solve the Test Cases while obtaining good performance indicators.

We had the same problems with the GoRoSoBo when we tried to solve the second part of the Test Cases (unscheduled offtake changes). The control problems for solving the second part were more important than the first part because the feedback algorithm does not have any notice about the disturbance until the water level at the checkpoints has been perturbed, that is, there is no possibility to change the gate trajectory to reduce the future water level errors before the disturbance perturbs the water level at the checkpoints. Instead, the scheduled offtake changes of the first part are known in advance by the feedforward algorithm at the beginning of the irrigation cycle.

It is absolutely necessary to introduce constraints in our optimization problem. The resulting constrained optimized problem is solved by the GoRoSoBo.

Before changing the code of our feedback algorithm to solve a constrained optimization problem, we studied other options. We tried to introduce “artificial constraints” in an unconstrained optimization problem with the aid of the Marquardt coefficient. Although the real objective of using the Marquardt coefficient is to improve an ill-conditioned matrix, we tried to use the Marquardt coefficient to restrict the gate movements. The Marquardt coefficient was computed with the assumption that the minimum gate position is equal to or greater than the lowest gate position allowed. The problem is that all the gate trajectories calculated by the algorithm are penalized for this Marquardt coefficient and the coefficient would only be able to restrict the minimum gate position. In some ways, applying the Marquardt coefficient is the same as modifying the influences of the elements of the HIM because we apply this coefficient at the diagonal of the Hessian matrix.

We dismissed this option and instead implemented in our algorithm a constrained optimization problem solved by the Lagrange–Newton method.

8.4. Algorithm controller from constrained optimization

As discussed in previous sections of this chapter, we need to include constraints on the gate movements. In case of disturbances, the feedback algorithm computes new gate trajectories, taking into account the constraints, to maintain the water level as close as possible to the target water level at the checkpoints, so the gate trajectories computed by the algorithm are physically possible and the best approximation to the optimal gate trajectories. Constraints have been established in a predictive control by many authors, for instance, by Soler [SOL 03] in GoRoSo (feedforward predictive control) or Zagona [ZAG 92] in MPC (model predictive control).

We can classify the different types of constraints as

– Overall constraints

– Functional constraints

Taking into account the experiences of these authors with constraints in predictive controllers, we introduced a set of overall constraints to avoid gate positions lower than the minimum gate opening and greater than the maximum gate opening (Figure 8.15):

$$U_{ij}(K) - U_{\min} \geq 0 \quad [8.16]$$

$$U_{\max} - U_{ij}(K) \geq 0, \quad [8.17]$$

where U_{min} is the minimum gate position, U_{max} is the maximum gate position, K is the number of the regulation period in a predictive horizon, "j" is the "jth" predictive horizon, and "i" is the "ith" gate. That is, $U_{ij}(K)$ is the gate position in the "jth" predictive horizon, for the "ith" gate at the regulation period K.

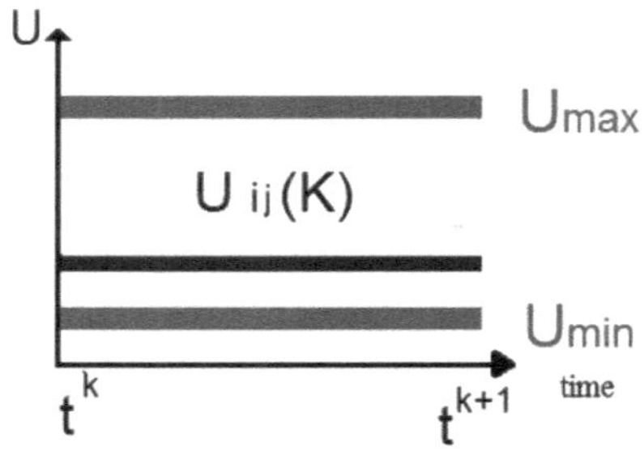

Figure 8.15. *Maximum and minimum gate position for the constrained optimization problem*

We also use functional constraints which restrict gate movements between time steps. The gate position between consecutive time steps should not be very different, because the free surface would change a lot between regulation periods and it would be difficult to control the canal. When does the control algorithm compute big changes in gate position for consecutive time steps?

This happens in the following situations:

– When significant flow changes are introduced into the canal.

– When small flow changes are introduced at particular points of the canal where the gates have a low influence on the target points.

It is possible that a gate has a low or null influence on several target points because these points are far away from this gate and the wave height generated by the gate is highly attenuated due to the friction forces, or because the wave generated by the gate arrives at the target points later than the end of the prediction horizon. If we want to fix the desired water level at the checkpoints with this gate, as the influence is low or null, the gate movement would be significant and functional constraints would be necessary (Figure 8.16):

$$\left|u_{ij}(K+1) - u_{ij}(K)\right| \leq dU\text{max}. \quad [8.18]$$

where $dU_{\max}$ is the maximum gate movement acceptable between consecutive regulation periods. This value can be associated with the characteristics of the servo-motors used to move the gate.

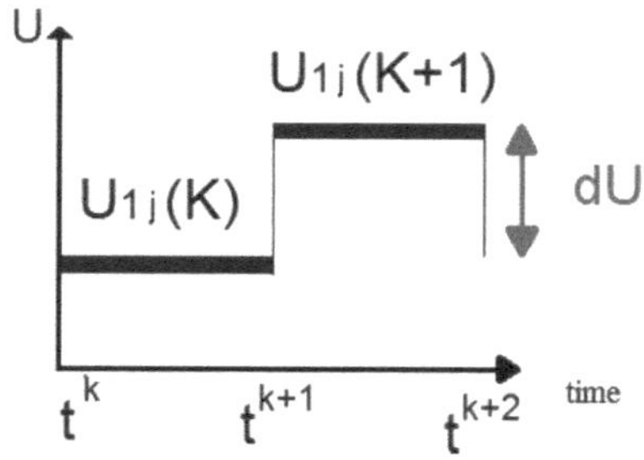

Figure 8.16. *Maximum gate movement between consecutive regulation periods for the constrained optimization problem*

After numerous tests using overall and functional constraints, the method did not work. All of these kinds of constraints were not enough. We thought that a new constraint was necessary to control the system because some gate movements destabilized the system.

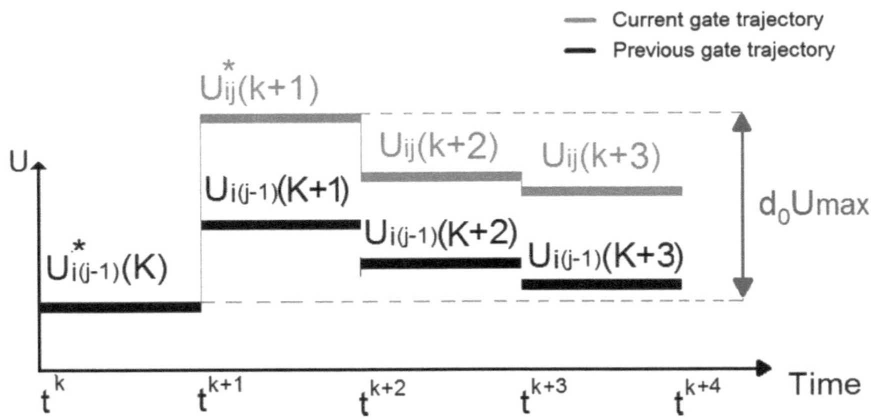

Figure 8.17. *Maximum gate movement between consecutive regulation periods for different predictive horizons for the constrained optimization problem. For a color version of the figure, please see www.iste.co.uk/bonet/canals.zip*

The predictive controller computes new gate trajectories for a predictive horizon at every regulation period, but only the gate positions during the first regulation period for the prediction horizon are sent to the actuator ($U^*_{i(j-1)}(K)$) (Figure 8.17). In the next regulation period, new gate trajectories will be computed by the control algorithm and only the gate positions at the first regulation period are sent to the actuator again ($U^*_{ij}(K)$). The gate trajectories used in the control structures contain the gate position in the first regulation period of each predictive horizon. For that reason, we need to restrict the gate movement between consecutive regulation periods of different predictive horizons (Figure 8.17). A feedback controller, such as the GoRoSoBo, needs new constraints between gate positions for consecutive regulation periods obtained in different predictive horizons. The equations associated with these new constraints are written as

$$d_0 U\max - \left| u_{ij}(K+1) - U_o \right| \geq 0$$
$$U_0 = u_{i(j-1)}(K), \qquad [8.19]$$

where d_0U_{max} is the maximum gate movement acceptable between consecutive regulation periods of different prediction horizons. U_0 is a vector with the gate positions at "*K*" regulation period for the prediction horizon (*j*-1), so U_0 is constant at the "*j*th" prediction horizon. The range of values to fix d_0U_{max} is similar to dU_{max} which could be estimated from empirical methods based on the experience of engineers in real cases, or from the features of the servo-motors which drive the gates.

As we introduced before, all optimization problems have to fulfill particular order conditions. To discuss the first- and second-order conditions, introduced in Appendix 2, which must be fulfilled by a constrained optimization problem, it is necessary to analyze the overall and functional constraint functions proposed in this constraint optimization problem. The Kuhn–Tucker conditions (first-order conditions) are necessary only if a particular condition is satisfied. That condition, called the constraint qualification, imposes certain restrictions on the constraint functions of a nonlinear programming problem, for the specific purpose of ruling out certain irregularities on the boundary of the feasible set, which would invalidate the Kuhn–Tucker conditions, as shown by Fletcher [FLE 87] and Luenberger [LUE 84]. The constraint functions must fulfill the constraint qualification and this condition is fulfilled in our optimization problem because the active constraint functions are linear and the first derivatives of these functions are linearly independent, see Soler [SOL 03]. We have to note that only active restrictions will be taken under consideration to evaluate the constraint qualification. A full discussion of the constraint qualification is given in Appendix 3.

8.4.1. *Optimized problem theory*

The methods used to solve the constrained optimization problem were two: the Lagrange–Newton method and the Active Set method.

The placing of one loop inside the body of another loop is called nesting. We use the Active Set method nested inside the body of the Lagrange–Newton method. These methods were chosen taking into account the good results obtained when they were implemented in the GoRoSo control algorithm by Soler [SOL 03].

The Lagrange–Newton method, also called the SQP (Sequential Quadratic Problem) method by Fletcher [FLE 87], is an iterative method, and each iteration sets a quadratic problem and every quadratic problem is solved by the Active Set method. We use these methods because they are robust and fast convergence methods in optimization problem. The SQP method is usually used in constraint optimization problems where the constraints are nonlinear. We can use this method even though the constraints introduced in our optimization problem are linear.

The SQP method is based on applying Newton's method to find the stationary point from the Lagrange function, hence its name: the Lagrange–Newton method, see Fletcher [FLE 87]. Both methods are used to solve the constrained optimization problem which is defined in each regulation period. A detailed explanation of the Lagrange–Newton and Active Set methods can be found in Appendix 4.

8.4.2. *Implemented algorithm*

The implemented algorithm consists of two iterative loops: an **interior loop** solves the quadratic problem by the Active Set method nested within the **external loop** which solves the SQP (Sequential Quadratic problem). The steps defined in the implemented algorithm are shown in the next flowchart (Figure 8.18) which is a type of diagram that represents every process, showing the steps as boxes, and their order by connecting them with arrows. As we did in the unconstrained optimization problem, we also added the Marquardt coefficient to the diagonal of the Hessian matrix to improve its condition number.

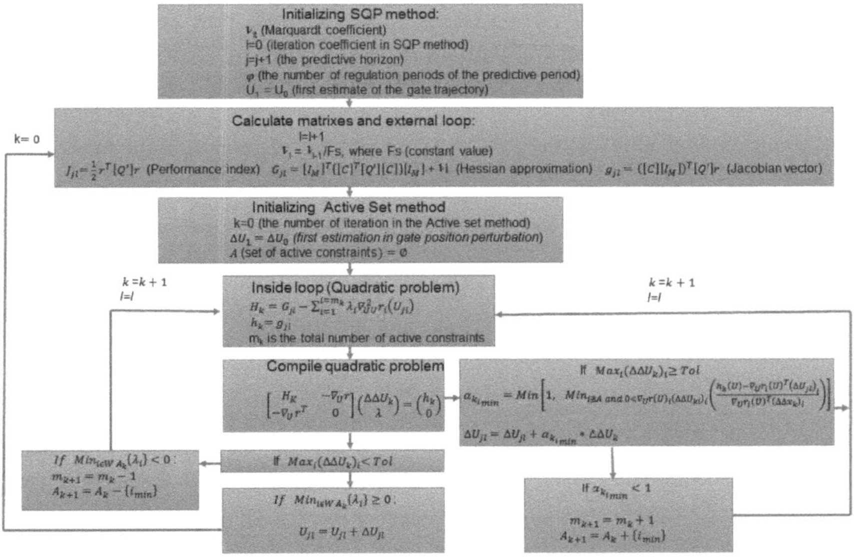

Figure 8.18. *GoRoSoBo scheduled: implemented version for the constrained optimization problem algorithm*

8.4.3. *Analysis of the future state computation*

If we want to accurately define the future canal state, we need to know the future disturbances. The problem is that we do not know the future changes in flow, because these changes can be caused by unscheduled flow extractions.

At this point, we have two options:

– Establish the current canal state without making future predictions.

– Establish the current canal state and make future predictions.

So the question, once the CSI has established the disturbances in the canal, would be: what is the future forecast of flow changes? Our future predictions could be that the disturbance will only exist during a regulation period, but we could also think that the disturbance will exist during several regulation periods. It is not an easy question because the control action calculated by the GoRoSoBo will be different in both options. In that sense, in the second option, the controller response will be stronger throughout the predictive horizon because the duration of the disturbance is longer than the first option. The canal state, that is, the velocity and the water levels along the canal, is different in both options (Figure 8.19) as well as the gate openings because the flow through the gates must be different, although the water level at the checkpoint (Y=Y*) is equal.

The problem is that the feedback controller modifies the gate opening according to the current canal's hydrodynamic state and the future predictions. In cases where we do not make future predictions about the disturbance, we suppose that the disturbance was introduced from the past to the present time, but if the disturbance has a long duration in the future, our future predictions will be wrong with respect to the reality. In that case, the feedback controller cannot calculate the optimum gate trajectories because it does not have a good prediction of the future canal state. In that sense, if our current predictions do not assume a disturbance with a long duration in the future, the gate trajectory solution is bad, because the water level error at the checkpoints is significant at the end of the simulation. The solution for gate trajectories obtained by the GoRoSoBo always maintains a stationary water level error at the end of the simulation, and that the sluice gate trajectory never corrects this error, see Figure 8.20. This is a generalized problem in all predictive controllers: nobody knows the value of the error and how it is distributed during the prediction horizon, see Figure 8.10.

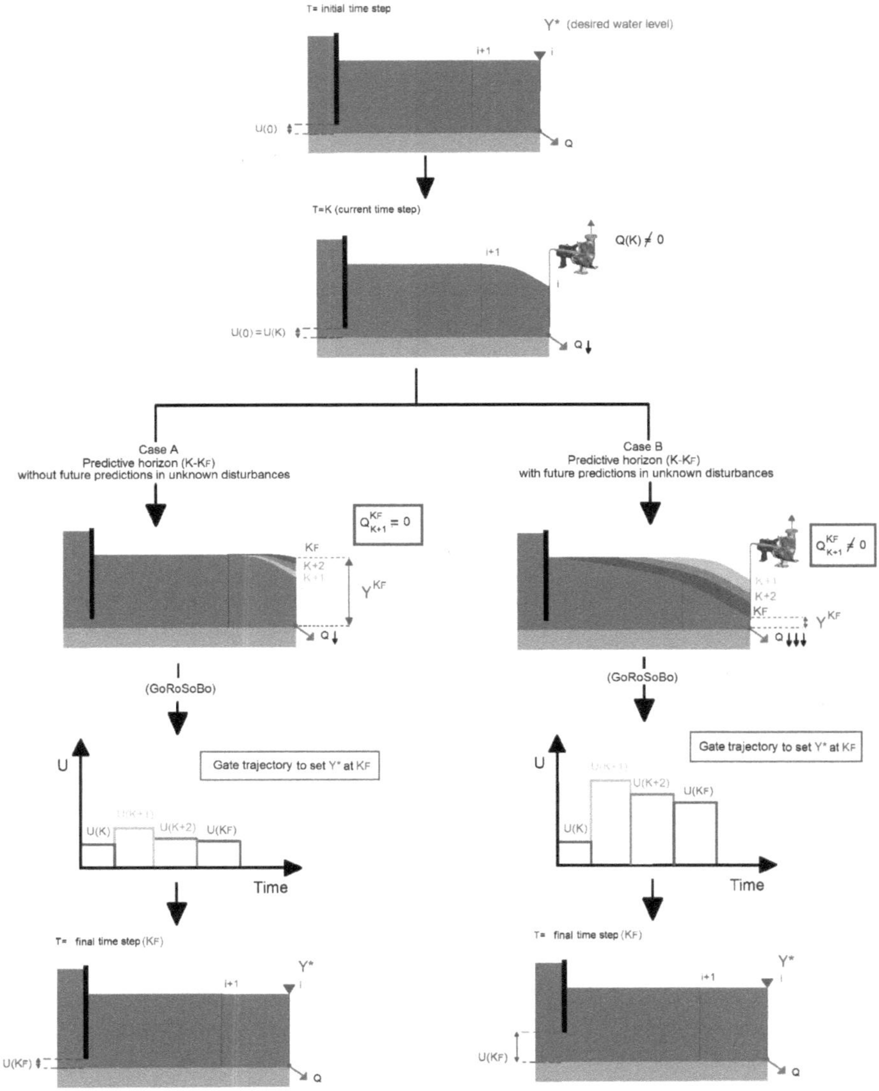

Figure 8.19. *Logic of control of the GoRoSoBo algorithm. For a color version of the figure, please see www.iste.co.uk/bonet/canals.zip*

There seems to not be a direct cause, but our predictions of the canal state in the future take more relevance depending on the kind of the disturbance introduced into the canal. After performing several tests with different kinds of disturbances using the GoRoSoBo, we concluded that if the disturbance is significant and especially for a "**long time duration**", although GoRoSoBo knows the current canal state, it is not enough to find the right solution of gate trajectories (Figure 8.20). The problems associated with this kind of disturbance are related to future predictions of the disturbance.

In summary, the feedback controller cannot manage the canal for that kind of disturbance when only considering the current canal state.

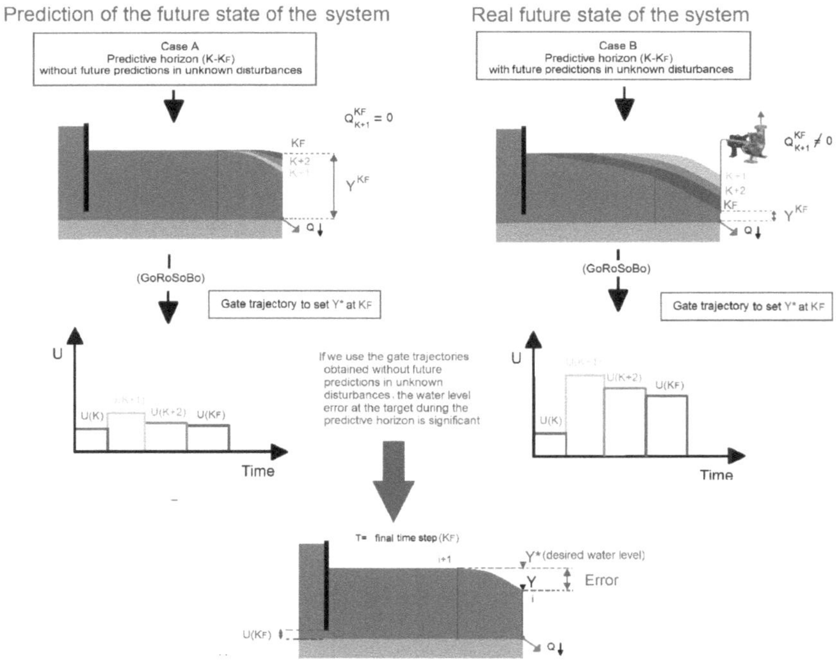

Figure 8.20. *Associated problems with disturbances with long durations in the GoRoSoBo without using future prediction. For a color version of the figure, please see www.iste.co.uk/bonet/canals.zip*

The effect of the disturbance along the canal during the predictive horizon is not easy to understand. In cases where the real disturbance is only introduced during a regulation time period, when the disturbance disappears, the water level is recovered quickly at checkpoints and the water level error in future will be low. In this way, the current changes in gate position to solve a minor error in the future would be small, as once the disturbance disappears, the water levels at checkpoints recover quickly by themselves, introducing little changes in gate trajectories. The algorithm will not introduce big perturbations in the system at the present time, taking into account that the error will reduce in the future and the perturbations introduced by the gate could be more important than perturbations for the disturbance.

But in cases in which the disturbance has a long time duration, it is necessary to establish a future prediction since the changes in gate trajectory introduced by the GoRoSoBo at the present time to correct the error in the future are fundamental. This is due to the control actions taken at the present time which are reflected in the future, see Figure 8.21. Failing to consider that the disturbance will continue over a long period of time is the cause of not driving the canal to the desired state and maintaining a stationary error between the desired and the real water level in the future.

For this reason, once the extracted flow vector is established by the CSI, the changes in flow at the present time will be the forecast for the predictive horizon in the GoRoSoBo (Chapters 10 and 11). In other words, the changes in the extracted flow vector obtained currently continue to be the changes in this vector during the prediction horizon. Although it is impossible to predict the future, there is no other way to drive the canal to the desired state in cases of this kind of disturbance; so we will make this future prediction.

It could be assumed that this solution probably is the best when the disturbance has a long time duration, but what will happen if the flow disturbance disappears in a short period of time, whereas our future predictions suppose the opposite? In such a case, the algorithm would calculate a gate trajectory taking into account a significant disturbance in the future. There is no simple answer to this question.

The result depends on the duration of the regulation period, as the CSI receives water level measurements every regulation period, and it recomputes the extracted flow vector every regulation period. If this period is short, the CSI determines sooner that the disturbance has disappeared and the GoRoSoBo would quickly recompute new gate trajectories to correct the last one. In the GoRoSoBo algorithm, the gate movements are restricted because we have tried to avoid instability in the numerical model. For these reasons, the changes in the water level introduced by excessive gate movements, due to wrong future predictions, are not so important.

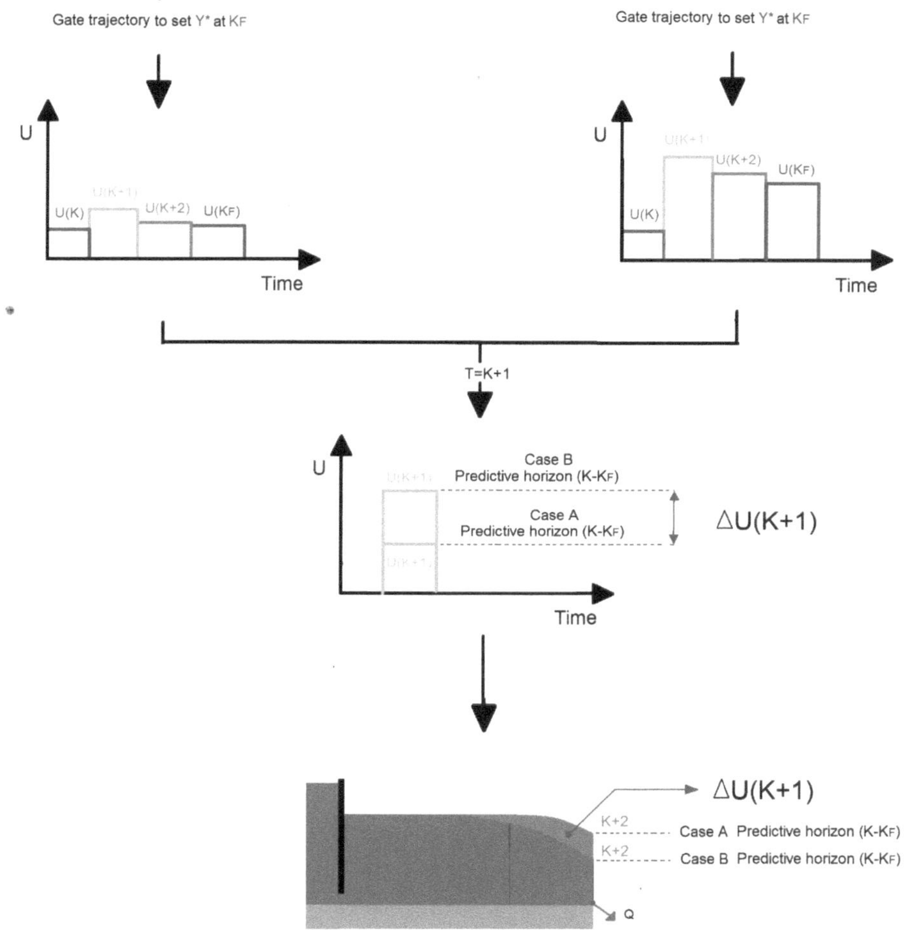

Figure 8.21. *Variation of the canal state taking into account different predictive horizons. For a color version of the figure, please see www.iste.co.uk/bonet/canals.zip*

8.4.4. *Conclusions of the constrained optimization problem*

In summary, we can conclude the following:

1) The GoRoSoBo obtains information from the CSI about the current canal state and the extracted flow, so the GoRoSoBo knows the water level error along the canal at the present time and it can make a prediction of the water level error for the

prediction horizon from this information. The only way to get the current state in a canal, without CSI, would be measuring every cross-section of the canal and using a calibrated predictive model, and the only way to establish the extracted flow in a canal would be to install flow meters at every extraction point.

2) The HIM, which represents the influence of a sluice gate on the canal state, is an excellent tool for the GoRoSoBo, which computes a new set of gate trajectories due to variations between the measured and the desired water level.

3) The HIM must be recomputed every regulation period in case of important disturbances.

4) Faced with significant disturbances, the GoRoSoBo should use constraints and mostly feedback controllers.

5) The Lagrange–Newton method used to solve a constrained optimization problem is a robust and fast convergence method.

6) GoRoSoBo is a feedback algorithm, so the gate trajectories are computed at every regulation period during the irrigation cycle. For that reason, the algorithm needs a new overall constraint to restrict the gate movements at consecutive regulation periods from different prediction horizons.

7) It is necessary that the GoRoSoBo introduces forecasted predictions in disturbances to drive the system to the desired state.

8) The GoRoSoBo algorithm shows promising results in all tests done, as we will show in Chapters 10 and 11.

8.5. Appendix I

8.5.1. *Unconstrained optimized problem (Levenberg–Manquardt)*

In the next equation, we show the "perturbation theory" by Brogan [BRO 85] developed (equation [8.20]) in a Taylor series around the gate reference trajectory:

$$Y = Y(U^* + \Delta U) = Y^*(U^*) + \left[\frac{\partial Y}{\partial U}\right]_{U^*} \cdot \Delta U + \frac{1}{2}\left[\frac{\partial^2 Y}{\partial U^2}\right]_{U^*} \cdot \Delta U^2 + \cdots \qquad [8.20]$$

Assuming that the disturbance caused is sufficiently small, we can neglect the terms after the first term (equation [8.21]), obtaining

$$\Delta Y = Y(U^* + \Delta U) - Y^* = \left[\frac{\partial Y}{\partial U}\right]_{u^*} \Delta U. \qquad [8.21]$$

On the other hand, it is possible to use the Hydraulic Influence Matrix (HIM), defined by Soler [SOL 03] as

$$\left[J^*\right]=\left[\frac{\partial Y}{\partial U}\right]_{U^*}. \qquad [8.22]$$

Taking into account all of these terms, we can write

$$\Delta U=\left[J^{*T}\cdot J^*\right]^{-1}\cdot\left[J^{*T}\right]\cdot\Delta Y. \qquad [8.23]$$

We can compute a set of ΔU gate movements to reach the desired water level ($Y=Y+\Delta Y$) at the checkpoints. The equation system can be solved by the Jacobi method, Cholesky method, singular value decomposition method or LU decomposition. There is an extensive bibliography about these methods, such as *Numerical Recipes in Fortran 77* [PRE 97a], where the Levenberg–Marquardt method is also illustrated.

The problem is to solve equation [8.23] at each time step. We can define the equation in a compact form:

$$[G]=\left[J^{*T}\right]\cdot\left[J^*\right] \qquad [8.24]$$

$$b'=\left[J^{*T}\right]\cdot\Delta Y. \qquad [8.25]$$

So replacing equations [8.24] and [8.25] in equation [8.23], we obtain

$$b'=[G]\cdot\Delta U. \qquad [8.26]$$

Although the system matrix is positive definite by definition, it is not well conditioned [SOL 03]. This implies that when solving the system, we can get numerical errors. To avoid this problem, we add a positive term to the diagonal of the Hessian matrix; this parameter is called Marquardt coefficient (ν). Much has been written on the theory of this coefficient, for instance, in Fletcher [FLE 87].

The system of equations can then be written as

$$b'=[G+\nu I]\cdot\Delta U, \qquad [8.27]$$

where [I] is the identity matrix of the same dimensions as $[G]$.

To solve the system (equation [8.27]), we have to make some assumptions. We have a diagonal matrix $[\Lambda]$ with the eigenvalues of the matrix $[G]$, and an orthonormal matrix ($[V]$: $[V\times V^T] = [I]$) defined by the eigenvectors associated with each eigenvalue of $[G]$. Multiplying both sides of equality (equation [8.27]) for the matrix $[V^T]$, we obtain

$$\left[V^T \cdot G + V^T \cdot \nu I\right]\left[V \cdot V^T\right]\Delta U = \left[V^T\right]\cdot b'. \qquad [8.28]$$

Knowing that the $[V]$ matrix is orthonormal, we can write the system as

$$\left[V^T \cdot G \cdot V + V^T \cdot \nu I \cdot V\right]\left[V^T\right]\Delta U = \left[V^T\right]\cdot b'. \qquad [8.29]$$

And according to the property $[\Lambda] = [V^T \times G \times V]$, it can be transformed (equation [8.29]) as follows:

$$\left[\Lambda + \nu I\right]\cdot\left[V^T\right]\cdot \Delta U = \left[V^T\right]\cdot b'. \qquad [8.30]$$

Furthermore, the fact of using orthonormal matrix $[V]$ allows us to ensure that

$$\left\|\left[\Lambda + \nu I\right]\cdot\left[V^T\right]\cdot \Delta U\right\| = \left\|\left[V^T\right]\cdot b'\right\|. \qquad [8.31]$$

To simplify (equation [8.30]), we can define the vectors Z and B as follows:

$$Z = \left[V^T\right]\cdot \Delta U \qquad [8.32]$$

$$B = \left[V^T\right]\cdot b' \qquad [8.33]$$

So Z is an unknown quantity and (equation [8.30]) can be transformed into

$$\left[\Lambda + \nu I\right] Z = B. \qquad [8.34]$$

From (equation [8.34]), the Z components can be separated easily because the new system matrix is now diagonal. Each of the Z components can be calculated as (equation [8.35])

$$Z(i) = \frac{B(i)}{\Lambda(i,i) + \nu}. \qquad [8.35]$$

Although the matrix [G] is positive definite, it is ill-conditioned too, and it is possible that some eigenvalues of [G] are negative, although very close to absolute zero. To prevent this problem, we introduced the Marquardt coefficient. If the Marquardt coefficient value is significant, ΔU will tend to zero and the water level in the targets cannot be recovered. However, if ν is very close to the minimum, you can get an excessively large ΔU, which causes large fluctuations in level. So you are looking for a range of values ν that provides a magnitude of a vector ΔU less than or equal to a certain value $\varepsilon_{\max}$. Depending on the value you chose for it, the solution will be different; for this reason, the value of this coefficient gets smaller in each iteration of the algorithm, in order to achieve a solution that is closer to the optimum solution.

Once you know the initial value for ν, the system (equation [8.35]) is resolved, as is the system (equation [8.32]), thus obtaining the desired values of ΔU (equation [8.36]):

$$\Delta U = [V] \cdot Z. \qquad [8.36]$$

In that sense, the algorithm computes a new gate trajectory ($U_a = U + \Delta U$) at time step K for a prediction horizon.

8.6. Appendix II

The Lagrange multipliers are the method used to solve the constrained optimization problem. The Lagrange multiplier method is explained by Fletcher [FLE 87], Luenberger [LUE 84] and Gill [GIL 81], although we will make a brief introduction to the method in this appendix. We must prove that our constrained optimization problem fulfils the order conditions, as we show in Appendix 2, to be sure that we obtain the optimal solution.

8.6.1. *First-order condition*

We could introduce the constrained optimization problem as follows:

$$\begin{aligned} &\text{minimize}\, J(U) = \frac{1}{2}\Big[[C]X_{ki}^{kF}(U) - Y^*\Big]^T [Q] \Big[[C]X_{ki}^{kF}(U) - Y^*\Big] \\ &r_k(U) = 0, \quad i \in I(U) \\ &r_k(U) \geq 0, \quad i \in NI(U), \end{aligned} \qquad [8.37]$$

where $J(U)$ is the objective function, $X_{ki}^{kF}(U)$ is the state vector from the regulation time step K_i to K_F, Y^* is the desired water level vector, Q is a weighting matrix, C is the discrete observer matrix and $r_k(U)$ is the "kth" constraint function, $I(U)$ is a set of equality constrains and $NI(U)$ is a set of inequality constraints.

Minimizing the objective function, we obtain the gate solution U^*. A feasible point U' fulfills all constraints, see Fletcher [FLE 87]. When a feasible point fulfills an inequality constraint as an equality, it is called active constraint. A set of constraints (equalities or inequalities) that are active in U' are called an active set.

We could define the constrained optimization problem from the Lagrange function as follows:

If U^* is a local minimum of the constrained optimization problem and λ_i are the Lagrange multipliers associated with the restrictions $r_k(U^*)$, we can establish the following system:

$$\mathcal{L}(U^*,\lambda^*) = J(U^*) - \sum_{i=1}^{nt} \lambda_k{}^* r_k(U^*). \qquad [8.38]$$

In that case, the first-order conditions or Khun-Tucker are written as

$$\begin{aligned}
&\nabla_U \mathcal{L}(U^*,\lambda^*) = 0\\
&r_k(U^*) = 0, \qquad i \in I\\
&r_k(U^*) \geq 0, \qquad i \in NI\\
&\lambda^*{}_k \geq 0, \qquad i \in NI\\
&\lambda^*{}_k r_k(U^*) = 0, \quad \forall i.
\end{aligned} \qquad [8.39]$$

If we develop the first derivative, the constraints and the multipliers, we will obtain

$$\begin{aligned}
\nabla_U \mathcal{L}(U^*) = (\nabla_U X_1^{K_F}(U^*)^T)\{[C]^T[Q]([C]X_1^{K_F}(U^*) - Y^*)\\
+ \nabla_x r^T \lambda^*\}
\end{aligned} \qquad [8.40]$$

$$+\nabla_U r^T \lambda^* = 0,$$

where

$$\lambda = (\lambda_1, \ldots, \lambda_{nt})^T$$

$$r = (r_1(X(U^*), U^*), \ldots, r_{nt}(X(U^*), U^*)).$$

If the solution U^* accomplishes this system (equation [8.40]), the solution fulfills the first-order condition.

So far, we have only considered the first-order conditions (that is, the first derivate). It is also possible to state the second-order conditions which give information about the curvature of the objective function at a local minimum.

8.6.2. *The second-order conditions*

In this section, we analyze the second-order condition, that is, we analyze the function in the neighborhood of the local solution, where $G(U^*)$ is a set of feasible directions. The full definition of this second-order condition is shown by Fletcher [FLE 87], where it is demonstrated how we reach the next expression:

$$s^T W(U, \lambda)^* \, s > 0 \qquad \forall s \in G(U^*) \tag{8.41}$$

$$W(U^*, \lambda^*) = \nabla^2_{UU} J(U^*) - \sum_{i=1}^{nt} \lambda_k \, \nabla^2_{UU} r_k(U^*) \tag{8.42}$$

where $W(U^*, \lambda^*)$ must be a definite positive matrix and "s" is a vector of any fixed length, the limiting vector "s" is referred to as a feasible direction to get U^*.

That means, the Lagrange function does not have a negative curvature for all feasible directions at U^*.

So we need to derive equation [8.40] to get $\nabla^2_{UU}\mathcal{L}$:

$$\begin{aligned}
\nabla^2_{UU}\mathcal{L}(U^*, \lambda^*) = {} & (\nabla_{UU}{}^2 X^{K_F}_{K_I+1}(U^*)^T)\{[C]^T[Q]([C]X^{K_F}_{K_I+1}(U^*) - Y^*) + \nabla_x r^T \lambda^*\} \\
& + [\nabla_U X^{K_F}_{K_{I+1}}(U^*)^T]\{[C]^T[Q][C] + [\nabla^r_{XX} r]\lambda\}[\nabla_U X^{K_F}_{K_{I+1}}(U^*)^T]^T \\
& + [\nabla_U X^{K_F}_{K_{I+1}}(U^*)^T] \textstyle\sum_{i=1}^{i=n_T} \lambda_i [\nabla^2_{UX} r_i] \\
& + [\nabla_U X^{K_F}_{K_{I+1}}(U^*)^T] \textstyle\sum_{i=1}^{i=n_T} \lambda_i [\nabla^2_{XU} r_i] + \sum_{i=1}^{i=n_T} \lambda_i [\nabla^2_{UU} r_i],
\end{aligned} \tag{8.43}$$

where $\nabla^2_{\lambda U}\mathcal{L} = \nabla^2_{U\lambda}\mathcal{L} = -\nabla_U r$.

As the constraint functions are linear, some terms of system [8.43] are null: $\nabla^2_{UX} r_i = 0, \nabla^2_{UU} r_i = 0,\ \nabla^2_{XX} r_i = 0$ and $\nabla^2_{U} r_i = 0$ for i=1,…,n_t, so:

$$\nabla^2_{UU}\mathcal{L}(U^*\lambda^*) = (\nabla_{UU}{}^2 X^{K_F}_{K_I+1}(U^*)^T)\{[C]^T[Q]([C]X^{K_F}_{K_I+1}(U^*) - Y^*)\} + [I_M]^T{}_Y[Q][C][I_M]_Y, \quad [8.44]$$

where

$$[I_M]_Y = [C](\nabla_U X^{K_F}_{K_I+1}(U^*)^T). \quad [8.45]$$

When the solution U^* is the optimal solution, so $([C]X^{K_F}_{K_I+1}(U^*)^T - Y^*)^T$ is close to zero. This means that the predicted behavior of the model and the desired behavior are very similar, so in this case, it is easy to demonstrate that $\nabla^2_{UU}\mathcal{L}(U^*,\lambda^*)$ becomes positive definite, as the $[C](\nabla_U X^{K_F}_{K_I+1}(U^*)^T)$ matrix is not singular. Moreover, if the lineal constraint functions verify the constraint qualification for any value of U, **then the $\nabla^2_{UU}\mathcal{L}(U^*)$ matrix is unfailingly positively defined for any values of U^*.**

Note: in this case, the Hessian matrix with linear constraints is equivalent to the Hessian matrix obtained from unconstrained optimization problems.

8.7. Appendix III

8.7.1. *Constraint qualification*

To verify the constraint qualification, it is obviously necessary to analyze the constraint functions defined in [8.16], [8.17], [8.18] and [8.19]:

$$r_{k-1}(U) = u_{ij}(K) - Umin = 0 \quad [8.46]$$

$$r_k(U) = Umax - u_{ij}(K) = 0 \quad [8.47]$$

$$r_{k+1}(U) = dU\max - \left(u_{ij}(K+1) - u_{ij}(K)\right) = 0 \quad [8.48]$$

$$r_{k+2}(U) = dU\max - \left(u_{ij}(K) - u_{ij}(K+1)\right) = 0 \quad [8.49]$$

$$r_{k+3}(U) = d_0U\max - \left(u_{ij}(K) - u_{(j-1)i}(K-1)\right) = 0 \quad [8.50]$$

$$r_{k+4}(U) = d_0U\max - \left(-u_{ij}(K) + u_{(j-1)i}(K-1)\right) = 0 \quad [8.51]$$

where $r_k(U)$ is the "*k*th" constraint function, *U*min and *U*max are the minimum and maximum gate positions allowed, d*U*max is the maximum gate movement between consecutive time steps and $d_0U$max is the maximum gate movement between consecutive time steps for different prediction horizons.

Before analyzing the constraint function, we have to point out that the theory for constrained optimization problems only considers the active constraints to calculate the solution *U*, and other restrictions are essentially ignored.

The first two overall constraints ([8.46] and [8.47]) exclude each other because, if a gate is totally open, constraint [8.47] is active; it cannot be totally closed, so constraint [8.46] cannot be active.

The functional restrictions [8.48] and [8.49] exclude each other too because these restrictions make reference to the gate movements between consecutive regulation time steps of a particular gate, and these constraints are completely opposed. The tendency of a gate movement should be to close or open, not both at the same time, so if constraint function [8.49] is active to restrict the maximum gate opening, constraint function [8.48] would not be active.

The overall constraints [8.50] and [8.51] are in the same situation as the last. These constraint functions are completely opposed. These constraints only can be active in the first regulation period for a prediction horizon.

In summary, you can ensure that the constraint qualification is verified for whatever combination of position values of the gates for two reasons: the first is that all constraints are linear and the second is that the active constraints are linearly independent.

8.8. Appendix IV

8.8.1. *Lagrange–Newton method*

We can define the constrained optimization problem as

$$\text{minimize}\, J(U) = \frac{1}{2}\left[[C]X_{ki}^{kF}(U) - Y^*\right]^T [Q]\left[[C]X_{ki}^{kF}(U) - Y^*\right]$$

$$\begin{aligned}
r_k(U) &= u_{ij}(K) - U\text{min} \geq 0 \\
r_k(U) &= U\text{max} - u_{ij}(K) \geq 0 \\
r_k(U) &= dU\text{max} - \left[u_{ij}(K+1) - u_{ij}(K)\right] \geq 0 \\
r_k(U) &= dU\text{max} - \left[u_{ij}(K) - u_{ij}(K+1)\right] \geq 0 \\
r_k(U) &= d_0U\text{max} - \left[u_{ij}(K) - u_{i(j-1)}(K-1)\right] \geq 0 \\
r_k(U) &= d_0U\text{max} - \left[u_{i(j-1)}(K-1) - u_{ij}(K)\right] \geq 0
\end{aligned} \quad [8.52]$$

where *U*min and *U*max are the minimum and maximum gate positions, *dU*max is the maximum gate movement between consecutive time steps and $d_0U$max is the maximum gate movement between consecutive time steps for different prediction horizons.

The Lagrange–Newton method solves a constrained optimization problem from the Lagrange function as follows:

$$\mathcal{L}(U,\lambda) = J(U) - \textstyle\sum_{i=1}^{nt} \lambda_k r_k(U), \quad [8.53]$$

where $r_k(U)$ is the "*k*th" constraint, λ_k is the "*k*th" Lagrange multiplier and n_t is the number of constraints.

We could establish the minimum of the Lagrange function as follows:

$$\nabla\mathcal{L}(U,\lambda) = 0. \quad [8.54]$$

Taking into account the Kuhn–Tucker conditions, it is possible to rewrite the optimization problem defined by the Lagrange function from the Taylor series for U, λ. The result of this method is a sequence of approximations U, λ [8.55] to get the solution vector U^* and the vector of optimum Lagrange multipliers λ^*:

$$\nabla\mathcal{L}(U+\Delta U, \lambda+\Delta\lambda) = \nabla\mathcal{L} + [\nabla^2\mathcal{L}]\begin{pmatrix}\Delta U\\ \Delta\lambda\end{pmatrix} + \cdots, \qquad [8.55]$$

where $\nabla\mathcal{L} = \nabla\mathcal{L}(U,\lambda)$. Neglecting higher-order terms and setting the left hand side to zero, that is, $\nabla\mathcal{L}(U+\Delta U,\lambda+\Delta\lambda) = 0$, because $U+\Delta U$ and $\lambda+\Delta\lambda$ are the solutions of the problem, it gives the iteration procedure:

$$[\nabla^2\mathcal{L}]\begin{pmatrix}\Delta U\\ \Delta\lambda\end{pmatrix} = -\nabla\mathcal{L}, \qquad [8.56]$$

where $\nabla^2\mathcal{L}$ is the Hessian matrix of the Lagrange function and $\nabla\mathcal{L}$ is the first derivative of the Lagrange function.

Where:

$$[\nabla^2\mathcal{L}] = \begin{bmatrix} W & -\nabla_U r\\ -\nabla_U r^T & 0\end{bmatrix} \qquad [8.57]$$

$$\mathrm{W} = \nabla^2_{UU}J(U) - \textstyle\sum_{i=1}^{nt}\lambda_i[\nabla^2_{UU}r_i] \qquad [8.58]$$

$$\nabla\mathcal{L} = \begin{pmatrix}\nabla_U J(U) - \sum_{i=1}^{nt}\lambda_i[\nabla_U r_i]\\ r\end{pmatrix}. \qquad [8.59]$$

So the system is written as follows:

$$\begin{bmatrix} W & -\nabla_U r\\ -\nabla_U r^T & 0\end{bmatrix}\begin{pmatrix}\Delta U\\ \Delta\lambda\end{pmatrix} = \begin{pmatrix}\nabla_U J(U) - \sum_{i=1}^{nt}\lambda_i[\nabla_U r_i]\\ r\end{pmatrix}. \qquad [8.60]$$

In fact, it is more convenient to write $\lambda^{K+1}=\lambda^K+\Delta\lambda$, and so to solve the equivalent system:

$$\begin{bmatrix} W & -\nabla_U r\\ -\nabla_U r^T & 0\end{bmatrix}\begin{pmatrix}\Delta U\\ \lambda\end{pmatrix} = \begin{pmatrix}\nabla_U J(U)\\ r\end{pmatrix}. \qquad [8.61]$$

Once the system is solved and the updated solution found, $U = U + \Delta U$ and λ, the solution must be checked to ensure it fulfills $\nabla\mathcal{L}(U,\lambda) = 0$; if this is not accomplished, the Hessian matrix and the gradient vector must be recalculated.

We can simplify matrix W because all constraints are linear in this problem, the element $\nabla^2_{UU} r_i = 0$, and the matrix W can be written in the form $W = \nabla^2_{UU} J(U)$ which is the Hessian matrix for unconstrained optimized problems.

System [8.61] has to be solved to get the solution ΔU. The problem is that we do not know initially which constraints are active, and it is not possible to limit δU. We cannot be sure that the solution $U^{k+1} = U^k + \Delta U$ is in a feasible region and the constraint optimization problem could became an unconstrained optimization problem. For this reason, we use the Active Set method to keep the solution in the feasible region defined by constraints.

8.8.2. *Active Set method*

The Active Set method was introduced to the algorithm code nested into the Lagrange–Newton method. The aim of the Active Set method is to compute a ΔU solution into a feasible region. The Active Set method never computes ΔU directly, but rather an approximation ($\Delta\Delta U$), where $\Delta U = \Delta U + \alpha \Delta\Delta U$. This ensures the method never computes a ΔU solution outside the feasible region. The active method is very useful when we want to solve a constrained optimization problem.

The idea underlying the Active Set method is to set inequality constraints in two groups which are treated as active, while all other variables are treated as inactive. The inactive constraints are essentially ignored, as the solution ΔU is not restricted by these constraints.

We can define the constrained optimization problem as follows:

$$\begin{aligned} &\text{minimize} \quad J(U + \delta U) \\ &r(U + \Delta U) \geq 0. \end{aligned} \qquad [8.62]$$

System [8.61] can be seen established as a quadratic problem, which consists of replacing the functions dependent on U by the respective linearization around U, establishing the following quadratic optimization problem:

$$\begin{aligned} J(U + \Delta U) &= J(U) + \nabla_U J(U)\Delta U + \frac{1}{2}\Delta U^T [\nabla^2_{UU} J(U)]\Delta U \\ r(U + \Delta U) &= r(U) + [\nabla_U r(U)]^T \Delta U. \end{aligned} \qquad [8.63]$$

If we establish the Lagrange function, we can write the system as follows:

$$\mathcal{L}(\Delta U, \lambda) = J(U) + \nabla_U J(U)\Delta U + \frac{1}{2}\Delta U^T[\nabla^2_{UU}J(U)]\Delta U$$
$$-(r(U) + [\nabla_U r(U)]^T\Delta U)\lambda. \qquad [8.64]$$

We must remember that $r(U) = 0$, because the solution (U) is always inside the feasible region, so only the active constraints are evaluated and these are only equality constraints.

If we establish that $\nabla\mathcal{L}(\Delta U, \lambda) = 0$, we can write the system as follows:

$$\nabla_U J(U) + [\nabla^2_{UU}J(U)]\Delta U - [\nabla_U r(U)]\lambda = 0$$
$$-[\nabla_U r(U)]^T\Delta U = 0. \qquad [8.65]$$

The system of equation [8.65] is nonlinear, for this reason we have to solve it iteratively as follows:

$$\begin{bmatrix} \nabla^2_{UU}J(U) & -\nabla_U r \\ -\nabla_U r^T & 0 \end{bmatrix}\begin{pmatrix} \Delta\Delta U \\ \Delta\lambda \end{pmatrix}$$
$$= -\begin{pmatrix} \nabla_U J(U) + +\nabla_U J(U)\Delta U + [\nabla_U r(U)]\lambda \\ 0 \end{pmatrix} \qquad [8.66]$$

$$\Delta U = \Delta U + \Delta\Delta U$$

$$\lambda = \lambda + \Delta\lambda.$$

In fact, it is more convenient to write $\lambda^{K+1}=\lambda^K+\Delta\lambda$, and so to solve the equivalent system:

$$\begin{bmatrix} \nabla^2_{UU}J(U) & -\nabla_U r \\ -\nabla_U r^T & 0 \end{bmatrix}\begin{pmatrix} \Delta\Delta U \\ \lambda \end{pmatrix} = -\begin{pmatrix} \nabla_U J(U) + \nabla_U J(U)\Delta U \\ 0 \end{pmatrix} \qquad [8.67]$$
$$\Delta U = \Delta U + \Delta\Delta U.$$

It is possible that, for any $\Delta U = \Delta U + \Delta\Delta U$, the solution $U = U + \Delta U$ is outside of the feasible set. For this reason, the parameter α is introduced

$$\Delta U = \Delta U + \alpha \Delta\Delta U$$
$$\alpha = \min\left[1, Min_{k \nexists A \text{ and } 0<\nabla_U r(U)\Delta\Delta U}\left(\frac{r_k(U) - \nabla_U r_k(U)^T \Delta U}{\nabla_U r_k(U)^T \Delta\Delta U}\right)\right]. \quad [8.68]$$

So that $\Delta U = \Delta U + \alpha \Delta\Delta U$. If $\alpha < 1$ then a new constraint becomes active, defined by the index which achieves the minimum in equation [8.68], and this index is added to the active set A. Then, if the other constraints are satisfied and the Lagrange multipliers turn out to be non-negative, the solution would be correct.

9

GoRoSoBo Application: Canal with Only One Pool

In this chapter, we present several numerical test cases on a canal with the GoRoSoBo algorithm, which works together with the CSI algorithm. GoRoSoBo is a feedback algorithm, which operates in real time with the objective to maintain the desired water level at checkpoints. In order to achieve this objective, the algorithm modifies the gate trajectories.

Under routine conditions, canal operations control the scheduled delivery at each control structure. If a small disturbance is introduced into the canal for a short period of time during the irrigation cycle, then the control strategy could be feedforward, as the flow is recovered quickly to the operative steady state. We are not concerned about those disturbances. However, when a significant disturbance (larger than 10% or 20% of the total flow rate) is introduced into the canal, we have to use a feedback strategy. The GoRoSoBo algorithm is tested in a canal with only one pool. This type of canal is chosen because the solution (gate trajectory) obtained by the algorithm is easier to understand and evaluate, and it can be used as a simple test to show the capacities of the proposed algorithm.

9.1. Introduction

In this chapter, we introduce different cases to test the GoRoSoBo. The objective is to check the accuracy of the control algorithm considering several types of disturbances with different magnitudes and durations under subcritical flow conditions.

The initial canal state is set from a known steady state, but during the irrigation cycle, we introduce scheduled deliveries and disturbances. The canal state will initially change to an unsteady state and later it will reach another steady state different from the original case. The control algorithm should maintain the water level at the checkpoint to fulfill the forecasted demand. These simple examples test the accuracy and the good behavior of the GoRoSoBo and the associated CSI algorithm working together.

9.2. Canal geometry

The geometry of the canal proposed in this chapter is based on Bautista's work [BAU 97]. This canal design was used by different authors, such as Whyle [WYL 69], Liu [LIU 92], Chevereau [CHE 91] and Soler [SOL 03], all of whom tested feedforward controllers. The canal geometry adopted in our examples is based on Liu's example as well as the scheduled demand, which was introduced in some cases of this chapter. There are also other authors who have tested their feedback controllers in canals with only one pool, similar to our canal, such as Durdu [DUR 09] and Litrico [LIT 06]. The canal geometry is shown in Figure 9.1.

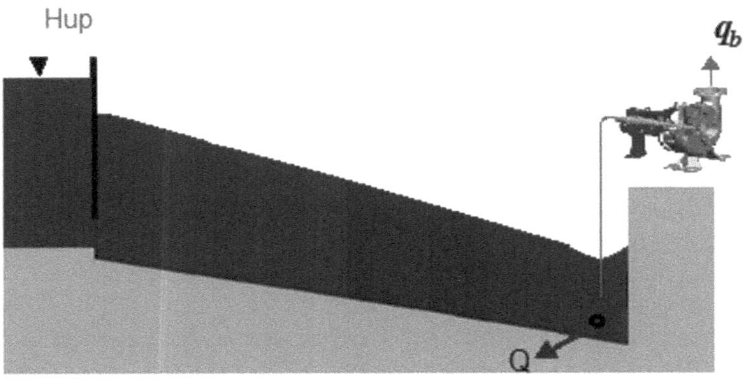

Figure 9.1. *Simple canal with only one pool*

9.3. Canal features

In accordance with the canal tested by Liu [LIU 92], the following are the features of this canal:

– Geometry and canal parameters: one pool of 2,500 m length and one sluice gate upstream of the canal (n_g=1); the bottom slope is 0.001, width is 5 m, bank slope (H:V) is 1,5:1 and the Manning coefficient is 0.025.

– Discretization: the spatial discretization of the canal is divided into cells of 25 m, so there are 101 nodes (n_s = 101). The regulation period is 5 min and the duration of the predictive horizon is 8 h, so there are 96 regulation periods (K_F = 96).

– Initial and boundary conditions: the upstream condition is fixed by a sluice gate that controls the water introduced into the canal from an upstream reservoir with a constant water level (3 m). Other boundary conditions depend on the flow extracted by the orifice offtake and the disturbance, which are introduced in the next section. At the end of the pool, there is an orifice offtake, a lateral spillway and a pump station. The steady state is the initial condition of the canal at the first regulation period. The flow rate in the canal is 5 m^3/s, and the desired water level at the checkpoint is 1.6 m (Figure 9.2).

Control structure	Numerical node upstream	Numerical node downstream	Gate discharge coefficient	Gate width (m)	Gate height (m)	Distance from the gate 1 (km)
0	0	1	0.61	5	2.0	0
1	101	101	–	–	–	2.5

Table 9.1. *Features of the canal control structure 1*

Control structure	Orifice offtake height (m)	Discharge coefficient/orifice offtake diameter (m)	Lateral spillway height (m)	Lateral spillway discharge coefficient	Lateral spillway width (m)
0	–	–	–	–	–
1	0.8	2/0.85	2.0	1.99	500

Table 9.2. *Features of the canal control structure 2*

There is a checkpoint at the last downstream node, and the water level is measured at the checkpoint in each regulation period.

The orifice offtakes are located downstream at the end of each pool, where the extracted flow through the offtake ($Q_{offtake}$) is described as follows:

$$Q_{\text{offtake}} = K\sqrt{y - y_0}$$
$$K = Cd \times \frac{\pi \times \emptyset^2}{4} \times \sqrt{2g}, \qquad [9.1]$$

where y is the water level in the canal at offtake, y_0 is the orifice offtake height equal to $y_{target}/2$, C_d is the discharge coefficient and Φ is the orifice offtake diameter.

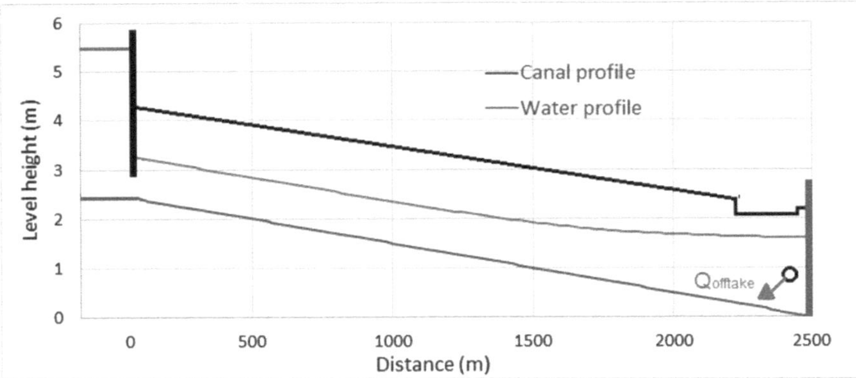

Figure 9.2. *Backwater curve of the canal (initial condition at the first regulation period of the irrigation cycle). For a color version of the figure, please see www.iste.co.uk/bonet/canals.zip*

9.4. Problem setting

We test several cases to check the algorithm's behavior under different disturbances.

There are two different scheduled deliveries:

– The forecasted demand is constant with a value of 5 m^3/s during the predictive horizon.

– The forecasted demand is variable in time, with a value of 5 m^3/s for the first two hours and 10 m^3/s for the next two hours; finally, the forecasted demand returns to 5 m^3/s during the last four hours of the test.

There are two different kinds of disturbances:

– The first one is called "variable disturbance", which is introduced 30 min after the test starts and lasts 1 h, with a value of 2 m^3/s.

– The second one is called the "step disturbance", which is introduced 30 min after the test starts and lasts until the end of the test, with a value of 2 m^3/s.

There are four other cases, which are the same as the previous ones; the only difference is the value of the disturbance (−2 m^3/s).

	Forecast demand	Disturbance
Case 1	Constant	"Step disturbance" (Figure 9.4)
Case 2	Constant	"Variable disturbance" (Figure 9.8)
Case 3	Variable	"Variable disturbance" (Figure 9.12)
Case 4	Variable	"Step disturbance" (Figure 9.15)
Case 5	Constant	"Negative step disturbance" (Figure 9.18)
Case 6	Constant	"Negative variable disturbance" (Figure 9.22)
Case 7	Variable	"Negative variable disturbance" (Figure 9.26)
Case 8	Variable	"Negative step disturbance" (Figure 9.30)

Table 9.3. *Scheduled and unscheduled deliveries introduced into the cases tested*

The cases, with a forecasted constant demand, are perfect to check how the algorithm is operating, as the only change in flow is due to the disturbance. It is easier to check the solution (the gate trajectory) obtained with the GoRoSoBo as well as how the water level at the checkpoints recovers to the desired value. Possibly, the easiest case to check the gate trajectory solution is test case 1.

Test case 2 produces two big flow changes in the canal: (1) when we introduce the "variable disturbance" and (2) when it disappears. The algorithm should readjust the gate position in a short period of time.

In test cases 3 and 4, the forecasted demand is not constant in time, which, in combination with the disturbances, tests the capacity of the GoRoSoBo algorithm to drive the canal to the desired state because the algorithm has to consider both flow changes.

In the other cases, we evaluate the same situation, but the disturbance is negative. Depending on the value and the duration of the forecasted demand and the disturbance, some test cases are more difficult to solve for the algorithm than others. For instance, in test case 7, the flow rate in the canal changes from 3 to 10 m^3/s in only 30 min, which is a big problem for the feedback algorithm.

9.5. Target value

The control algorithm should establish the target value for the canal, maintaining a certain water level at checkpoint, which is 1.6 m. This water-level target must be constant for the irrigation cycle. Figure 9.3 shows the desired water-level vector (dots) defined at the checkpoint in every regulation period in a computational grid.

The duration of the test is 8 h (K_F = 8 h). There is only one checkpoint downstream of the pool, so the number of water-level targets is 96 ($n_T = n_t{*}\ K_F$).

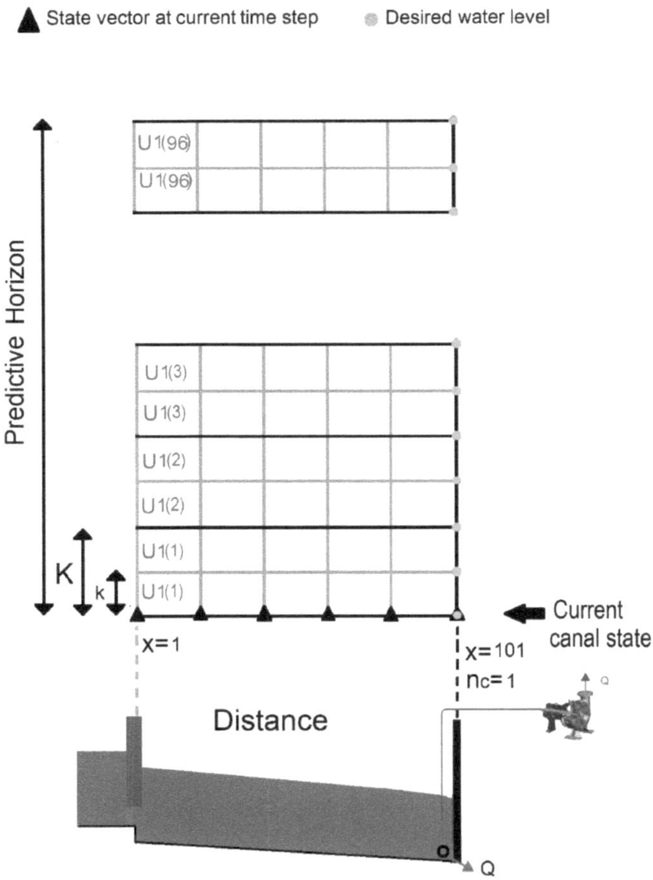

Figure 9.3. *Time discretization of a test in the canal*

9.6. Constraints and trajectories

The sluice gate opening must not be greater or smaller than U_{max} or U_{min}, respectively, and the gate movements between successive regulation periods (dU_{max}) should be physically acceptable, as described in Chapter 8.

Taking into account the results obtained by Soler [SOL 03] in the analysis of constraints in these problems, the overall and functional constraints are presented in Table 9.4. The constraints are determined as a certain percentage of the gate height and are imposed onto the gate position.

U_{min} (%)	U_{max} (%)	dU_{max} (%)
0.05	90	2.5

Table 9.4. *Overall and functional constraint values*

9.7. Test results

We represent the test results in several graphs. The first graph represents the scheduled delivery and the disturbance during the irrigation cycle; the second graph shows the disturbance calculated by the CSI algorithm; the third one shows the gate trajectory in each regulation period during the irrigation cycle and the fourth graph shows the water level at the checkpoint during the irrigation cycle.

We only show the disturbance calculated by the CSI algorithm in cases where the disturbance is different from the others, because there are cases where we use the same type of disturbance. If the CSI algorithm can estimate the disturbance in one of them, then it can estimate it in others, too.

In the first graph of each case, the dashed line represents the scheduled delivery and the solid line represents the disturbance. In the third and fourth graphs of each case, the dashed line represents the results obtained using a feedforward algorithm (GoRoSo) and the solid line shows the results obtained by the GoRoSoBo algorithm. The sluice gate trajectories obtained by GoRoSoBo and GoRoSo can be compared to obtain accurate results.

$$\text{Gate position} = \frac{\text{Gate opening}}{\text{Gate height}} \qquad [9.2]$$

The gate trajectory is defined by the gate position in time, and the gate position is defined as follows:

Case 1

In the following graph, we show the scheduled (dashed) and unscheduled (solid) deliveries introduced in this case.

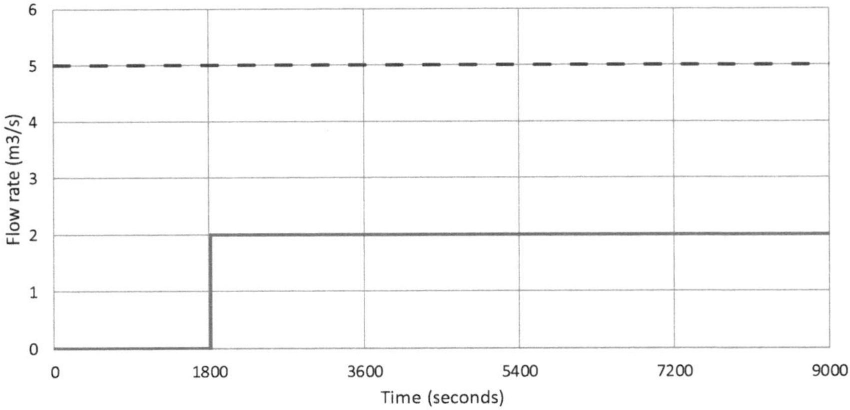

Figure 9.4. *Scheduled and unscheduled deliveries introduced in the system (Case 1)*

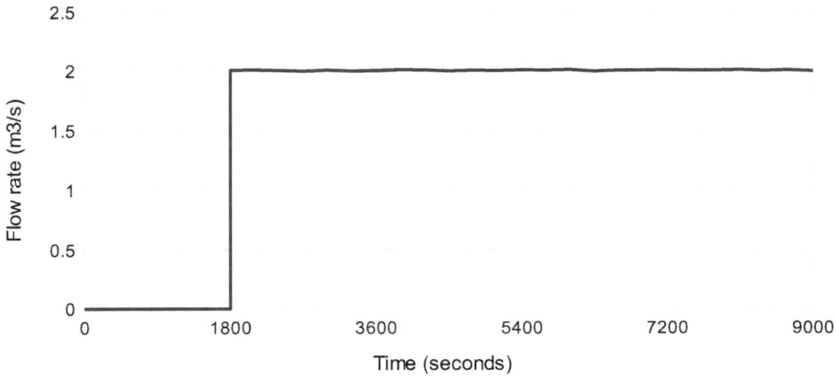

Figure 9.5. *Disturbance calculated by the CSI (Case 1)*

At the beginning of the test, the flow in the canal is steady, the scheduled delivery is constant for the whole demand period and the sluice gate position remains fixed. After the first 30 min (T = 1,800 s), a disturbance is introduced into the system, although the algorithm has no notice until the next regulation period

(T = 2,100 s) once the water level is measured at the checkpoint, and the water level has been already reduced by more than 10 cm, from 1.60 to 1.50 m (Figure 9.7).

Once the water-level measure is sent to the CSI algorithm, it calculates the extracted flow vector for a past time horizon; thus, the algorithm establishes the disturbance introduced in the system (Figure 9.5). All of this information is sent to the GoRoSoBo, which modifies the sluice gate trajectory to keep the water level constant at the checkpoint, which will not increase until three regulation periods later (T = 2,700 s), because once the sluice gate generates a wave to increase the water level at the checkpoint, the wave has to travel to the checkpoint (that is, there is a time delay). Once the wave generated by the sluice gate arrives at the checkpoint, the water level increases, quickly recovering the target water level (1.60 m) at T = 3,700 s (Figure 9.7).

The maximum deviation between the measured and desired water levels is around 27.5 cm, from 1.6 to 1.325 m at 2,700 s, so the sluice gate movement must be quite important to reduce this deviation quickly (Figure 9.6). Gate position changes from 0.125 to 0.34 during a regulation period (between T = 1,800 s and T = 2,100 s).

However, the feedforward algorithm does not modify the gate trajectory during the irrigation cycle because the scheduled demand is constant, and the algorithm does not have any notice about the introduction of a disturbance in the canal and the decrease of water level at the checkpoint. The water level decreases to 1.1 m at the orifice offtake, so the flow extracted by the offtake is only 3 m^3/s (Figure 9.7). The total extracted flow is still 5 m^3/s: 3 m^3/s extracted by the offtake and 2 m^3/s by the disturbance. The difference in gate opening between the feedforward and feedback algorithms at T = 9,000 s is due to the disturbance. A difference of 0.06 in gate opening represents a flow of 2 m^3/s in this canal under these flow conditions.

Once the algorithm recovers the desired water level, the gate movement should be zero in this test. Because the water level is close to the desired value, there exists a water level deviation of about ±1 mm. This deviation between both water levels is due to residual errors in the numerical computations of both algorithms (CSI and GoRoSoBo), and hence, the gate movements are not zero.

Case 1 is the easiest case for the GoRoSoBo algorithm and the most suitable to check the algorithm's behavior because the demand deliveries and the disturbance are constant in time. As can be seen from Figure 9.7, the control algorithm keeps the water level constant at the checkpoint (1.6 m), and in only 25 min, it recovers a water level error of 27.5 cm at the checkpoint.

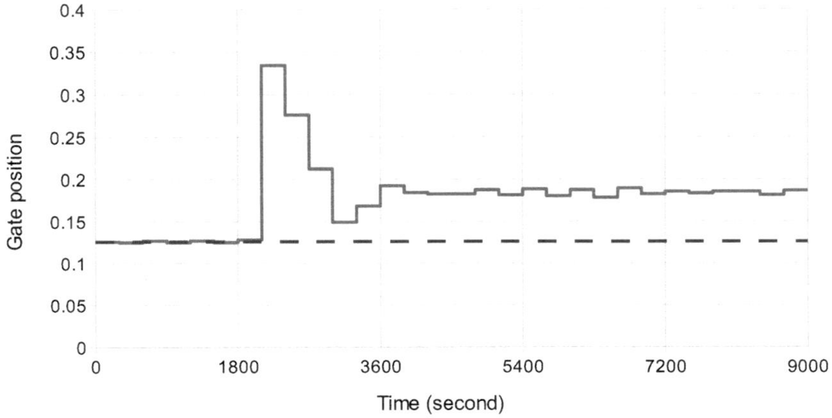

Figure 9.6. *Gate trajectory obtained by GoRoSoBo (solid line)/ GoRoSo (dashed line) in Case 1. For a color version of the figure, please see www.iste.co.uk/bonet/canals.zip*

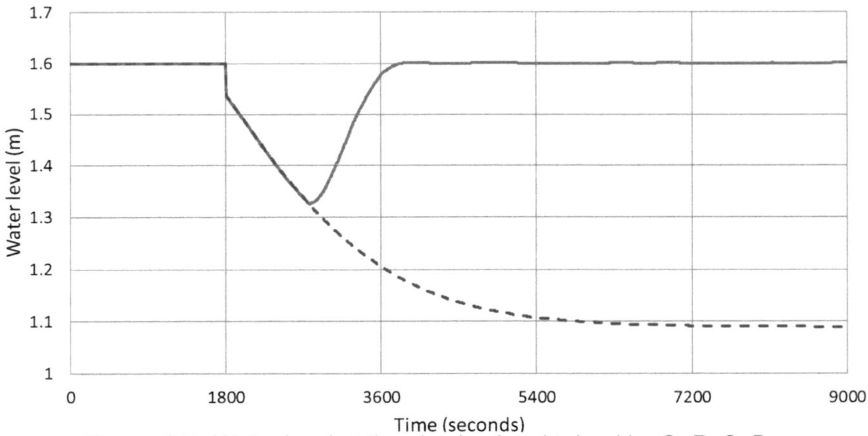

Figure 9.7. *Water level at the checkpoint obtained by GoRoSoBo (solid line)/GoRoSo (dashed line) in Case 1. For a color version of the figure, please see www.iste.co.uk/bonet/canals.zip*

Case 2

In the following graph, we show the scheduled (dashed) and unscheduled (solid) deliveries introduced in this case.

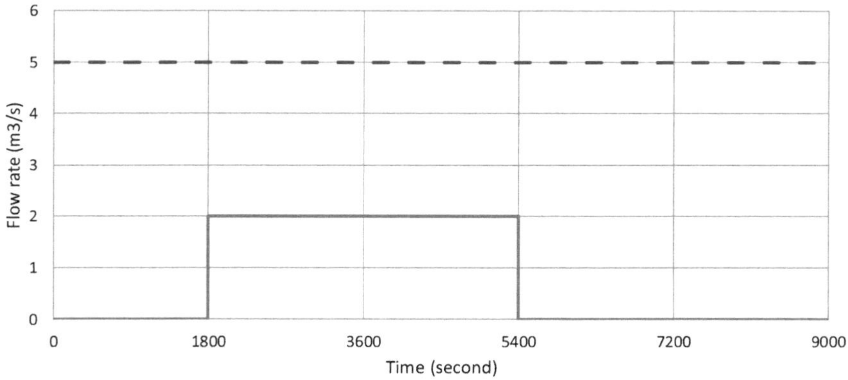

Figure 9.8. *Scheduled and unscheduled deliveries introduced in the system (Case 2)*

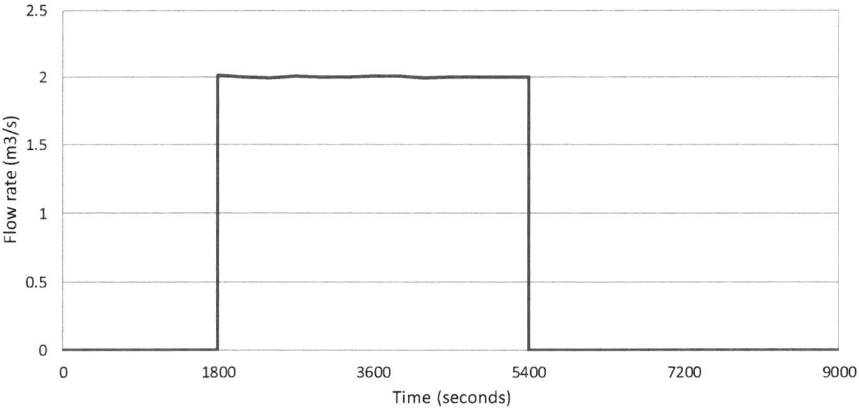

Figure 9.9. *Disturbance calculated by CSI (Case 2)*

This case is similar to the last case only for the first 5,400 s because we introduce a disturbance that only lasts 1 h, as it is variable in time. The water level at the checkpoint decreases from 1.60 to 1.32 cm (28 cm) in just 15 min as in the last test (see Figure 9.11).

When a disturbance is introduced into the canal (Figure 9.8), the sensor measures the water level at the checkpoint and these data are sent to the CSI algorithm, which calculates the disturbances considering the scheduled demand and the gate trajectory in the past time horizon (Figure 9.9).

All these data are sent to the GoRoSoBo, which calculates a gate position of about 0.34 at T = 2,100 s (Figure 9.10) to increase the water level at the checkpoint. Once the disturbance disappears at T = 5,400 s, the water level at the checkpoint increases significantly (Figure 9.11) and the CSI algorithm calculates the new disturbances and sends the information to the GoRoSoBo, which also calculates a new gate trajectory. The gate closes to reduce the flow rate introduced from the reservoir to the canal, and the gate position is reduced from 0.34 to 0.0015 at T = 5,700 s (Figure 9.10). The gate position returns to 0.125 at T = 7,500 s because the flow conditions of the system return to the initial conditions.

However, the feedforward algorithm does not modify the gate trajectory during the irrigation cycle because the algorithm establishes an optimal gate trajectory at the beginning of the irrigation cycle, considering only the scheduled deliveries. As there is no feedback between the feedforward controller and the canal state, the algorithm does not know that the water level is decreasing at the checkpoint. The water level decreases until the flow extracted by the offtake is close to 3 m^3/s (Figure 9.11). Once the disturbance disappears, the water level recovers its target value because the flow conditions are equal to the initial conditions. In this regard, if the disturbance is constant, as in the last test, the feedforward algorithm is not able to return to the target value; therefore, the farmers only obtain 3 m^3/s from the orifice offtake during the irrigation cycle.

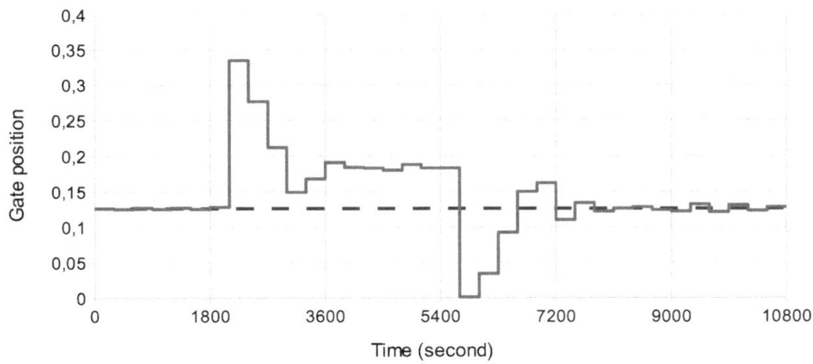

Figure 9.10. *Gate trajectory obtained by GoRoSoBo (solid line)/GoRoSo (dashed line) in Case 2. For a color version of the figure, please see www.iste.co.uk/bonet/canals.zip*

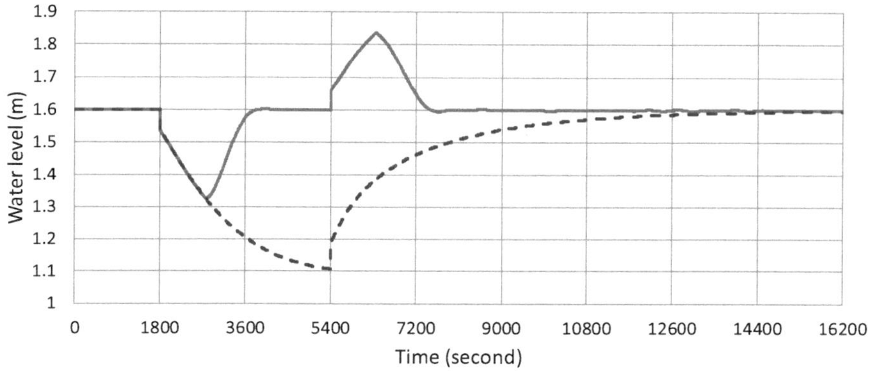

Figure 9.11. *Water level at the checkpoint obtained by GoRoSoBo (solid line)/GoRoSo (dashed line) in Case 2. For a color version of the figure, please see www.iste.co.uk/bonet/canals.zip*

Case 3

In the following graph, we show the scheduled (dashed) and unscheduled (solid) deliveries introduced in this case.

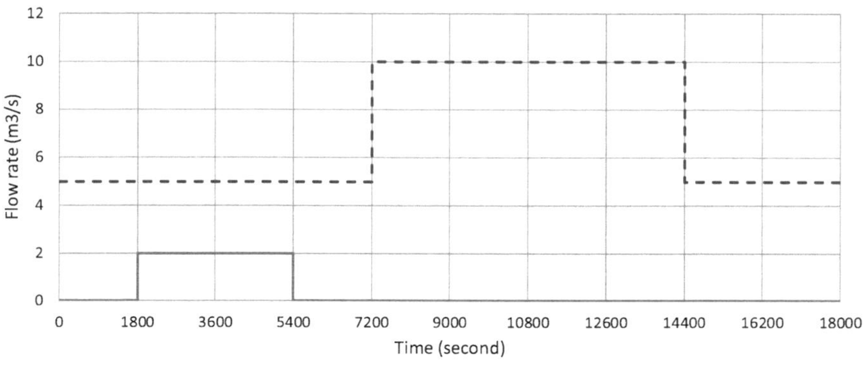

Figure 9.12. *Scheduled and unscheduled deliveries introduced in the system (Case 3). For a color version of the figure, please see www.iste.co.uk/bonet/canals.zip*

This case is similar to the last case only for the first two hours, so we will obtain the same sluice gate trajectory to reduce the deviation between the measured and desired water levels.

As shown in the previous cases, the CSI algorithm is able to calculate the disturbance introduced into the system and these data are sent to the GoRoSoBo, which calculates a new gate position at T = 2,100 s to drive the system to the desired state (Figure 9.13).

The feedforward algorithm does not change the gate trajectory once the disturbance is introduced into the canal and the water level at the checkpoint decreases to 1.1 m at T = 5,400 s. The water level starts to increase when the disturbance disappears at T = 5,400 s. Figure 9.12 shows that the scheduled delivery is not constant in time. For this reason, 30 min before changing the forecasted demand from 5 to 10 m^3/s, the feedforward algorithm modifies the gate trajectory to reduce the future deviation between the measured and desired water levels (Figure 9.13). The free surface has to be prepared for an important change in water demand, as the forecasted demand changes from 5 to 10 m^3/s. As this change in flow has been scheduled, the feedforward algorithm opens the gate to provide more flow to the canal. As more disturbances are introduced into the canal during the irrigation cycle, the water level reaches the target value at T = 10,800 s.

In a feedback strategy, the GoRoSoBo should consider the scheduled demands in addition to the disturbance. The GoRoSoBo should close the gate at T = 5,700 s (see Figure 9.13) because the disturbance has disappeared at T = 5,400 s, but at the same time, it should open the gate because the forecasted demand will change from 5 to 10 m^3/s at T = 7,200 s. The algorithm must take into account all these flow changes and maintain the desired water level (1.6 m) at the checkpoint (Figure 9.14).

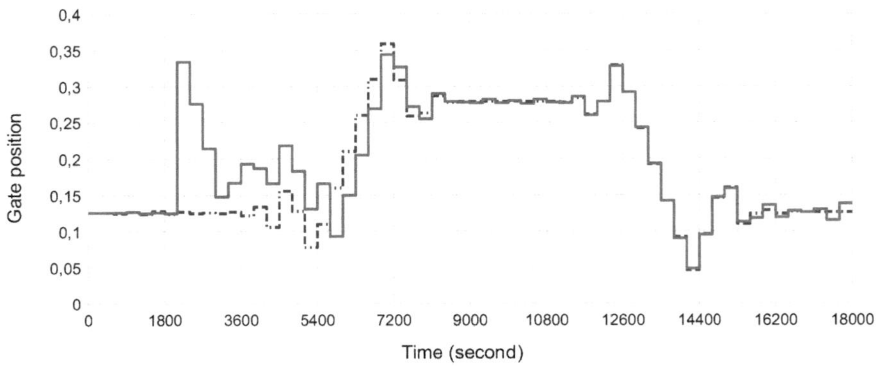

Figure 9.13. *Gate trajectory obtained by GoRoSoBo (solid line)/GoRoSo (dashed line) in Case 3. For a color version of the figure, please see www.iste.co.uk/bonet/canals.zip*

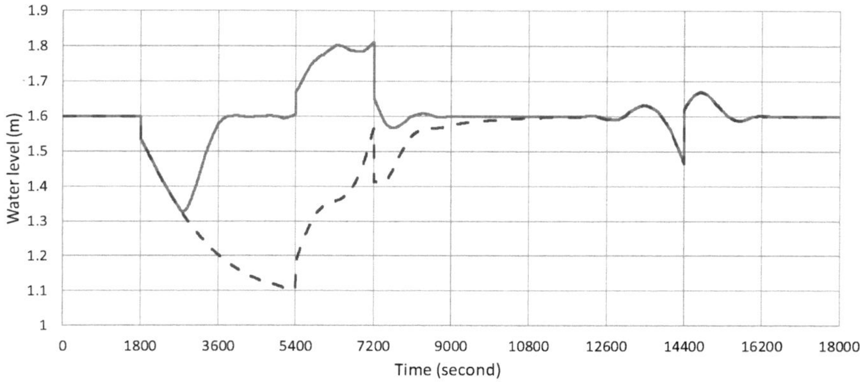

Figure 9.14. *Water level at the checkpoint obtained by GoRoSoBo (solid line)/GoRoSo (dashed line) in Case 3. For a color version of the figure, please see www.iste.co.uk/bonet/canals.zip*

Case 4

In the following graph, we show the scheduled (dashed) and unscheduled (solid) deliveries introduced in this case.

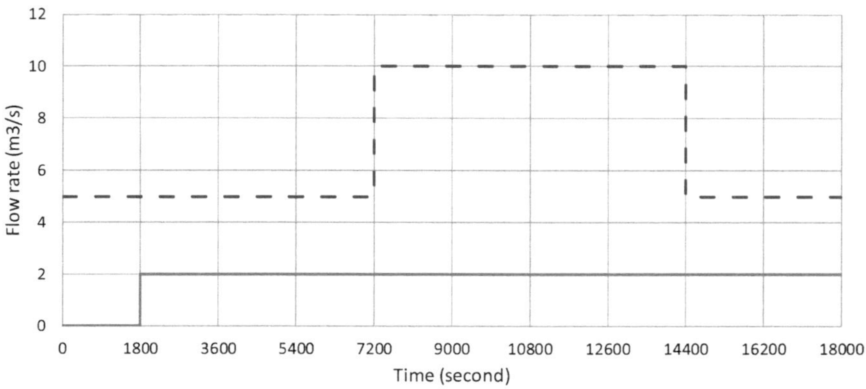

Figure 9.15. *Scheduled and unscheduled deliveries introduced in the system (Case 4). For a color version of the figure, please see www.iste.co.uk/bonet/canals.zip*

Here, the disturbance is constant in time for the first 30 min and the forecasted demand is variable (Figure 9.15). The gate trajectory of the GoRoSo is proportional to the gate trajectory of the GoRoSoBo after the second hour (Figure 9.16), and the difference of both trajectories is due to the constant disturbance of 2 m^3/s. Both water levels at the checkpoint are proportional to the third hour (see Figure 9.17).

The feedforward algorithm does not change the gate trajectory when the disturbance is introduced into the canal at T = 1,800 s and the water level decreases to 1.1 m, because the disturbance does not disappear until the end of the test. There is an increase in water demand from 5 to 10 m^3/s, which is considered by the feedforward algorithm because it is scheduled, but the water level never recovers the target value because the disturbance does not disappear during the irrigation cycle.

Figure 9.16 shows that the difference in gate position of 0.06 between the feedforward and feedback algorithms represents a flow of 2 m^3/s in this canal under these flow conditions. In this regard, a difference in water level of 0.5 m at the checkpoint also represents a flow of 2 m^3/s (Figure 9.17).

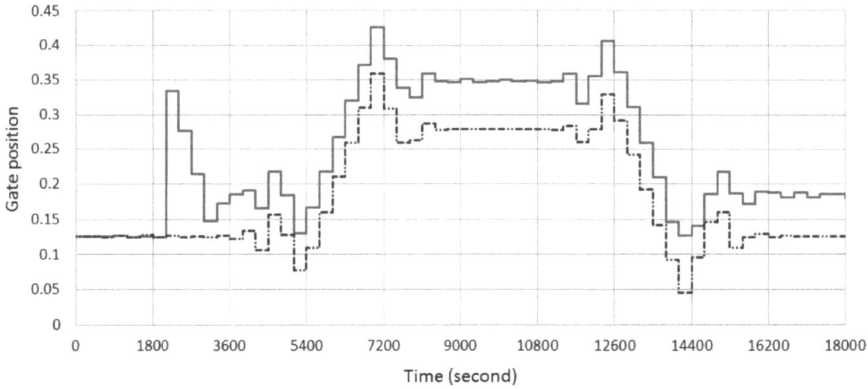

Figure 9.16. *Gate trajectory obtained by GoRoSoBo (solid line)/GoRoSo (dashed line) in Case 4. For a color version of the figure, please see www.iste.co.uk/bonet/canals.zip*

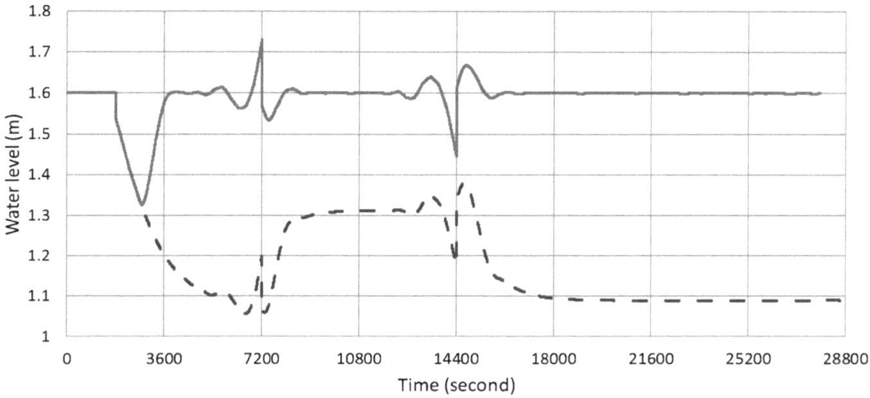

Figure 9.17. *Water level at the checkpoint obtained by GoRoSoBo (solid line)/GoRoSo (dashed line) in Case 4. For a color version of the figure, please see www.iste.co.uk/bonet/canals.zip*

Case 5

In the following graph, we show the scheduled (dashed) and unscheduled (solid) deliveries introduced in this case.

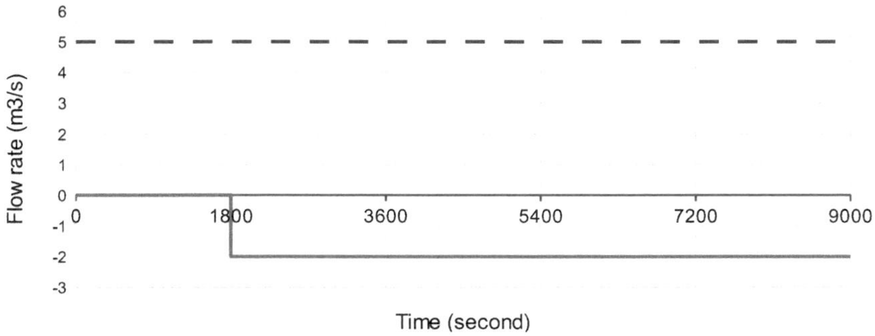

Figure 9.18. *Scheduled and unscheduled deliveries introduced in the system (Case 5). For a color version of the figure, please see www.iste.co.uk/bonet/canals.zip*

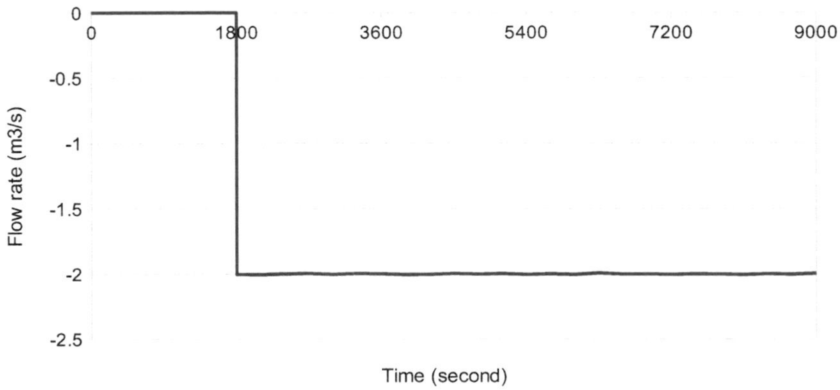

Figure 9.19. *Disturbance calculated by CSI (Case 5)*

This case is the same as the first one. The only difference between them is that the disturbance has a negative value (Figure 9.18); therefore, it is not an output flow, but an input flow to the canal.

Figure 9.49 shows that the CSI algorithm can also estimate the disturbance when it has a negative value. This information is sent to the GoRoSoBo, which modifies the sluice gate trajectory to keep the water level constant at the checkpoint, which will not decrease until three regulation periods later at T = 2,700 s (Figure 9.21).

The gate trajectories obtained in Case 5 are the opposite to those obtained in Case 1 (see Figures 9.20 and 9.6, respectively). In this case, the gate position between T = 2,100 s and T = 3,000 s is the minimum allowed (U_{min}), so the flow through the gate is the minimum flow allowed to reduce the water level at checkpoint. The water-level trajectory at the checkpoint is also opposite to that in Case 1 (see Figures 9.21 and 9.7, respectively).

As the feedforward algorithm does not have any information about the negative disturbance, it cannot calculate a new gate trajectory and the water level at the checkpoint increases over the canal, so there is canal overflow at T = 4,614 s. As the disturbance does not disappear during the irrigation cycle, the lateral spillway operates until the end of the test.

The difference in gate position between the feedforward and feedback algorithms at T = 9,000 s is due to the negative disturbance. A difference of 0.06 in gate position represents a flow of −2 m^3/s in the canal, and as the disturbance has the same absolute value as in Case 1, the difference in gate position is the same.

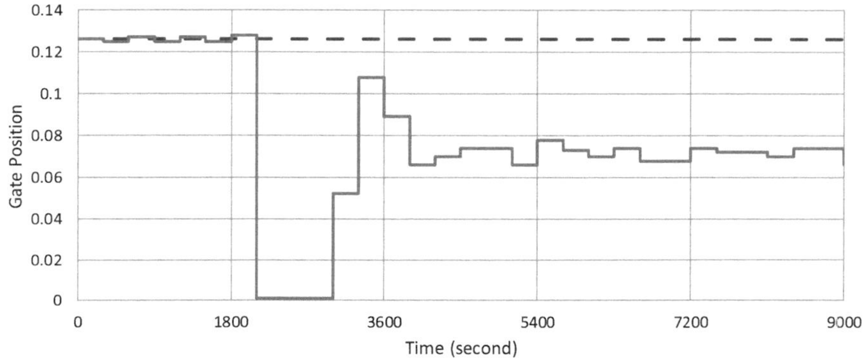

Figure 9.20. *Gate trajectory obtained by GoRoSoBo (solid line)/GoRoSo (dashed line) in Case 5. For a color version of the figure, please see www.iste.co.uk/bonet/canals.zip*

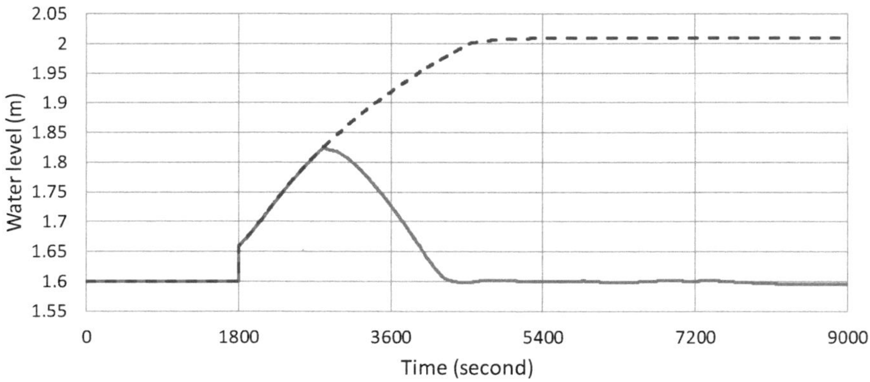

Figure 9.21. *Water level at the checkpoint obtained by GoRoSoBo (solid line)/GoRoSo (dashed line) in Case 5. For a color version of the figure, please see www.iste.co.uk/bonet/canals.zip*

Case 6

In the following graph, we show the scheduled (dashed) and unscheduled (solid) deliveries introduced in this case.

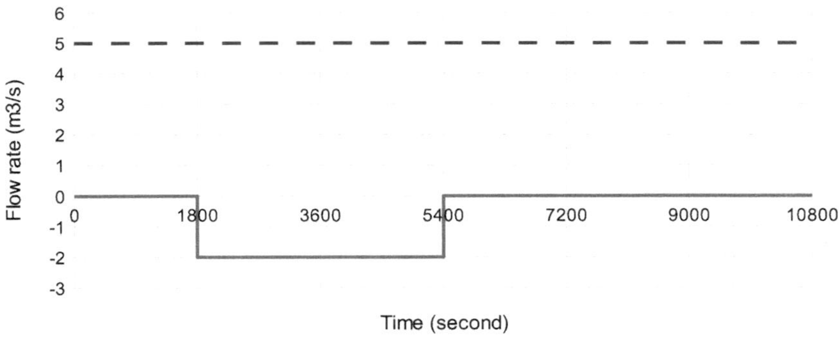

Figure 9.22. *Scheduled and unscheduled deliveries introduced in the system (Case 6). For a color version of the figure, please see www.iste.co.uk/bonet/canals.zip*

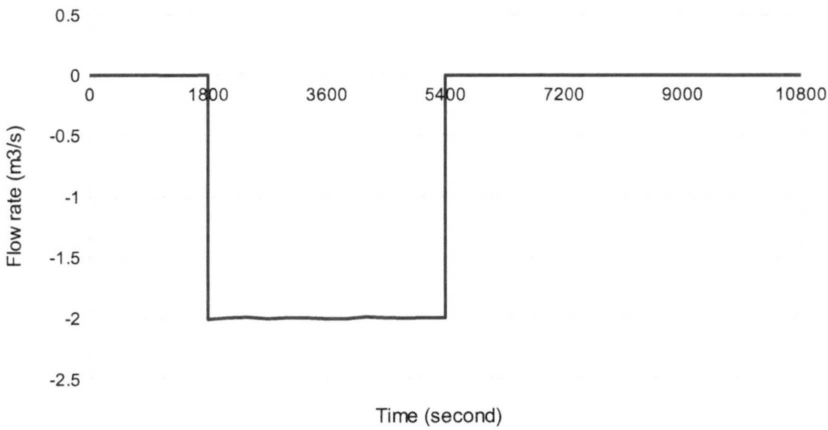

Figure 9.23. *Disturbance calculated by CSI (Case 6)*

This case is the same as the second case; the only difference between them is that the disturbance has a negative value (Figure 9.22).

Figure 9.23 shows that the CSI algorithm can also estimate the disturbance when it has a negative and variable value. The water level increases at T = 1,800 s due to the negative disturbance, and the GoRoSoBo closes the gate to reduce the water level at the checkpoint, when the disturbance disappears at T = 5,400 s; thus, GoRoSoBo should open the gate because the flow through the gate is too small to deliver 5 m^3/s through the offtake.

The gate trajectories obtained in Case 6 are opposite to those obtained in Case 2 (see Figures 9.24 and 9.10, respectively). In this case, the gate position reaches the minimum allowed (U_{min}) between T = 2,100 s and T = 3,000 s. For this reason, this gate trajectory is not equal to the inverse gate trajectory of Case 2. The same analogy could be made about the water-level trajectories at the checkpoint in both cases.

As the feedforward algorithm does not have any information about the negative variable disturbance, there is no change in gate position; therefore, water level at the checkpoint increases to 2 m at T = 4,614 s (Figure 9.25), there is a canal overflow, the lateral spillway operates between T = 4,614 s and T = 5,400 s, when the disturbance disappears and the water level decreases at the checkpoint.

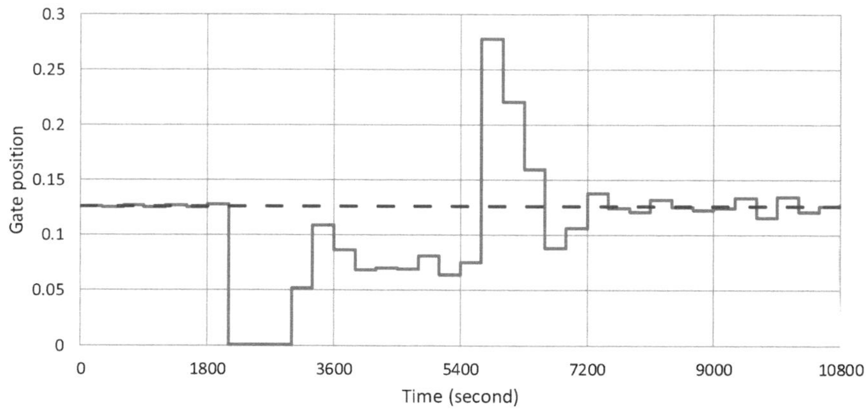

Figure 9.24. *Gate trajectory obtained by GoRoSoBo (solid line)/GoRoSo (dashed line) in Case 6. For a color version of the figure, please see www.iste.co.uk/bonet/canals.zip*

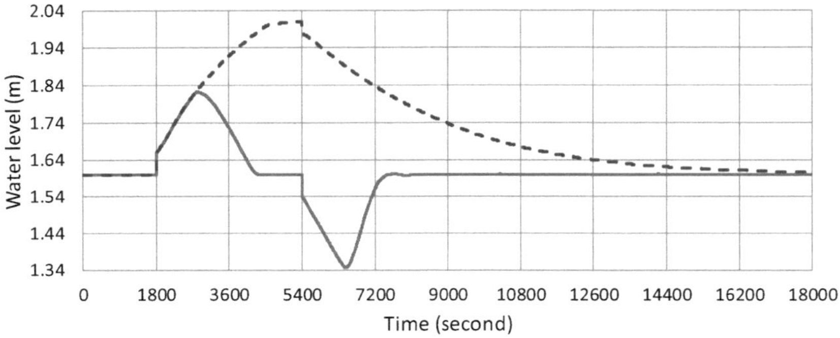

Figure 9.25. *Water level at the checkpoint obtained by GoRoSoBo (solid line)/GoRoSo (dashed line) in Case 6. For a color version of the figure, please see www.iste.co.uk/bonet/canals.zip*

Case 7

In the following graph, we show the scheduled (dashed) and unscheduled (solid) deliveries introduced in this case.

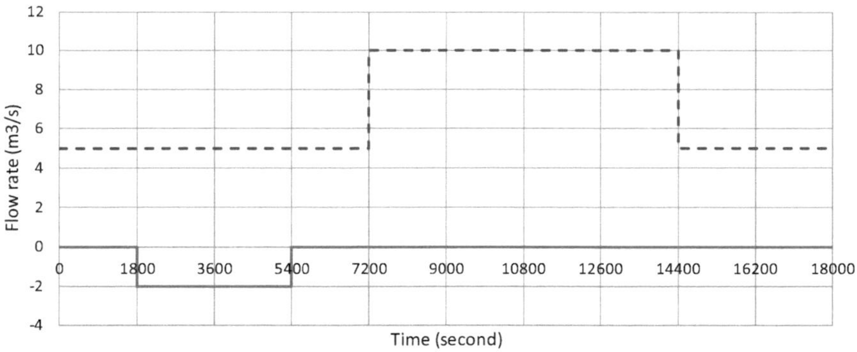

Figure 9.26. *Scheduled and unscheduled deliveries introduced in the system (Case 7). For a color version of the figure, please see www.iste.co.uk/bonet/canals.zip*

This case is the same as the third case, the only difference is that the disturbance is negative (Figure 9.26). The gate trajectories are similar in the feedback and feedforward strategies after the first two hours (see Figures 9.27 and 9.13), because the disturbance disappears at T = 5,400 s and the only flow change is due to the forecasted demand.

Figure 9.28 shows that the water level at the checkpoint obtained using the feedforward algorithm is not recovered until T = 12,000 s, that is, 6,600 s after the disturbance disappears, whereas, the feedback algorithm recovers the water level at the checkpoint before the disturbance disappears at T = 4,000 s. It should be highlighted that the enclosed area (A) by the water level trajectory (Figure 9.29) from 1,800 to 4,000 s is similar to the enclosed area (B) from 5,400 to 6,900 s. We can conclude that the volume of water delivered to the farmers from 1,800 to 4,000 s (where the water level is over the target value (1.6 m)) is similar to the volume of water not delivered from 5,400 to 6,900 s; therefore, the volume of water delivered to the farmers during the irrigation cycle is similar to the volume demanded by the farmers.

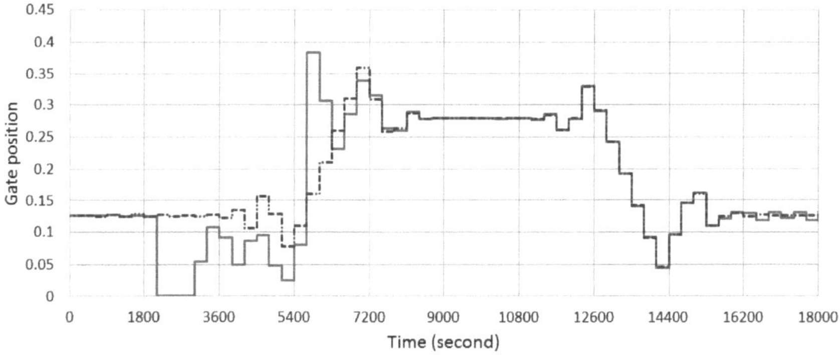

Figure 9.27. *Gate trajectory obtained by GoRoSoBo (orange continuous line)/GoRoSo (blue discontinuous line) in Case 7. For a color version of the figure, please see www.iste.co.uk/bonet/canals.zip*

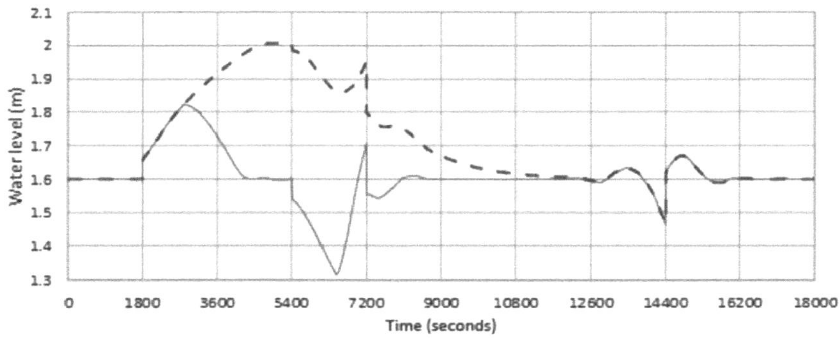

Figure 9.28. *Water level at the checkpoint obtained by GoRoSoBo (orange continuous line)/GoRoSo (blue discontinuous line) in Case 7. For a color version of the figure, please see www.iste.co.uk/bonet/canals.zip*

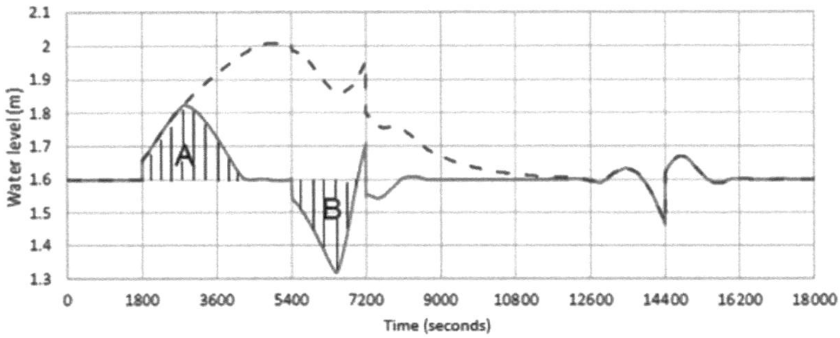

Figure 9.29. *Enclosed area versus water level at the checkpoint obtained by GoRoSoBo (solid line) in Case 7. For a color version of the figure, please see www.iste.co.uk/bonet/canals.zip*

Case 8

In the following graph, we show the scheduled (dashed) and unscheduled (solid) deliveries introduced in this case.

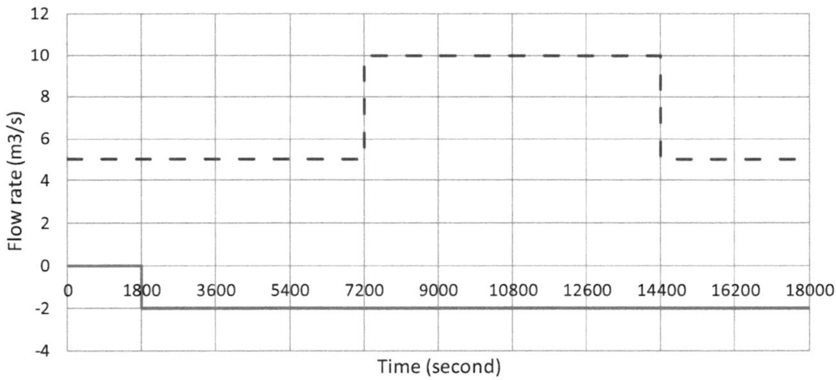

Figure 9.30. *Scheduled and unscheduled deliveries introduced in the system (Case 8). For a color version of the figure, please see www.iste.co.uk/bonet/canals.zip*

This case is the same as the fourth case; the only difference is that the disturbance has a negative value (Figure 9.30). The gate trajectories are completely different in both cases for the first 5,400 s, and then they are proportional until the end of the test (see Figures 9.31 and 9.16). The difference of both gate trajectories is due to the constant disturbance that increases the total flow rate by 2 m^3/s in this case and reduces the total flow rate by 2 m^3/s in Case 4.

Figure 9.31 shows that the feedback and feedforward algorithms have equivalent gate positions between T = 3,900 s and T = 18,000 s. The feedback algorithm calculates a gate position of 0.072 at T = 18,000 s, whereas the feedforward algorithm calculates a gate position of 0.126 at T = 18,000 s. The feedforward algorithm does not know that there is an inlet flow of 2 m^3/s in the canal, and the water level is 0.409 m over the target value (1.6 m); therefore, the flow delivered by the offtake is 6.147 m^3/s when the scheduled demand is 5 m^3/s.

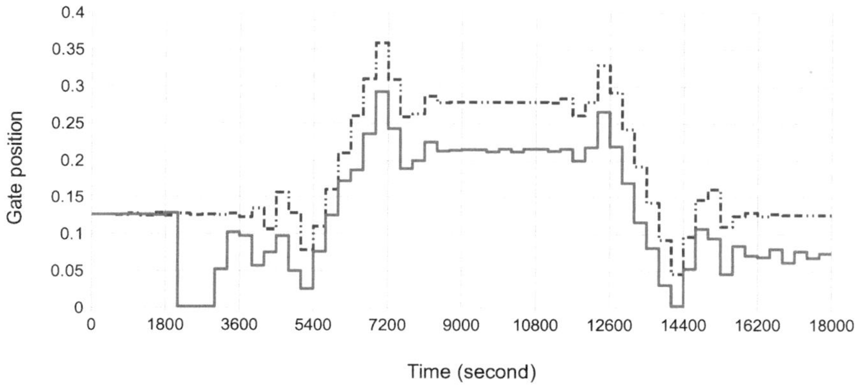

Figure 9.31. *Gate trajectory obtained by GoRoSoBo (solid line)/GoRoSo (dashed line) in Case 8. For a color version of the figure, please see www.iste.co.uk/bonet/canals.zip*

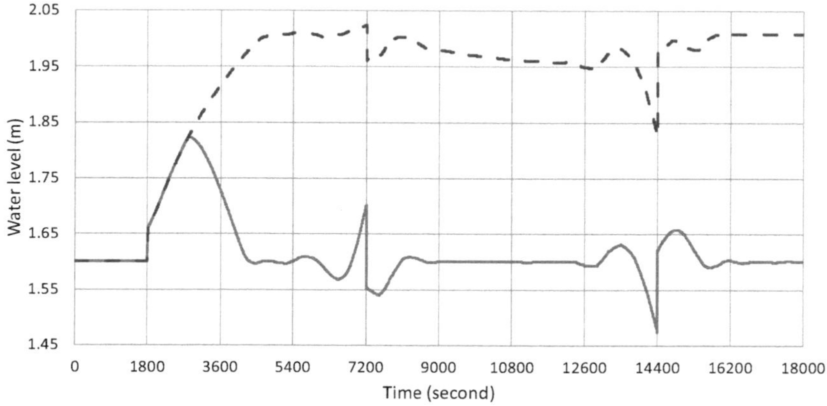

Figure 9.32. *Water level at the checkpoint obtained by GoRoSoBo (solid line)/GoRoSo (dashed line) in Case 8. For a color version of the figure, please see www.iste.co.uk/bonet/canals.zip*

The last four test cases can be compared with the first four. In cases where the forecasted demand is constant, the gate trajectories are opposite, depending on whether the disturbance is negative or positive.

9.8. Conclusion

Once the results of the different cases have been shown, the following conclusions can be drawn:

1) Some test cases, introduced in this chapter, are difficult to solve by a control algorithm, because the flow changes in forecasted demand and disturbance are 100% and 40%, respectively, over the total flow rate in the canal.

2) The algorithm computes the best sluice gate trajectory to maintain the water level at the checkpoint. In this regard, the algorithm maintains the forecasted demand at the delivery point, although a disturbance perturbs the free surface.

3) To control the canal state in real time, the algorithm must calculate a new gate trajectory, in the case that a disturbance modifies the water level at the checkpoint from the target. The changes in gate trajectories will be more important depending on the magnitude of this disturbance. For this reason, it is necessary to restrict the sluice gate movements, which could be excessive in these cases.

4) The CSI and GoRoSoBo algorithms operate together in all the test cases. The water-level measurements, scheduled demand and gate trajectory are sent to CSI if a disturbance is introduced into the system, to establish this disturbance. All these data are sent to the GoRoSoBo, which calculates the best sluice gate trajectory to reach the desired behavior at the checkpoint.

In the next chapter, more examples will be introduced to test the GoRoSoBo algorithm.

10

ASCE Test Cases

In this chapter, we show the application of the control algorithm in the test cases proposed by the ASCE [CLE 98]. The ASCE Task Committee on Canal Automation Algorithms proposed test cases on two real canals for the comparison of control algorithms. The test cases cover a set of conditions that may be found in practice. Two test schedules were adopted for each canal: in each of the four cases, scheduled offtake changes are introduced during the first part and unscheduled offtake changes, which are unknown to the algorithm, are introduced in the second part. In this chapter, as we want to evaluate GoRoSoBo, we only test the second part of every case.

The two canals proposed by the ASCE were chosen because they are completely different. The Corning canal is large and has a mild slope, whereas the lateral canal WM in Maricopa Stanfield district is small and steep with short response times. The aim in these tests is to check the robustness and accuracy of the control algorithm for different canals. The response of the algorithm should be good in both cases.

Two of the test cases proposed by the ASCE for feedback controllers introduce huge changes to unscheduled flow deliveries in a short period of time. The aim is to check the capacity of the feedback controller to drive the canal state and remove the disturbance effect caused by the unscheduled flow change. The test schedule strategies in both canals are similar, starting from a steady state. The canal conveys a high flow rate to deliver water to the farmers due to an important demand delivery. Two hours after the beginning of the irrigation cycle, the demand delivery decreases sharply because the water demand is reduced to a minimum flow rate in the whole canal without the knowledge of the watermaster. These two test cases are quite restrictive for most of the feedback controllers because the changes in water demand are very important; for instance, in the Corning canal, there are unscheduled offtake changes of around 4 m^3/s. In the lateral canal WM, as it is a steep canal characterized by little storage, feedback controllers have even more problems in controlling the state of the canal due to its small storage capacity and the important changes in water demand.

The lateral outlet flow in the test is produced by orifice offtakes at the downstream end of each pool and by pump stations.

Detailed canal features are introduced in this chapter as well as the unscheduled and the scheduled deliveries for each test. The behavior of GoRoSoBo is evaluated in each test and the results are shown in the next section and compared to the results obtained by other feedback control algorithms, such as CLIS or Pilote. More information about the CLIS and Pilote control algorithms can be found in Liu *et al.* [LIU 98] and Malaterre *et al.* [MAL 95].

10.1. Canal features

Test canal 1 is based on lateral canal WM in the Maricopa Stanfield Irrigation and Drainage District in central Arizona. Details on canal 1 can be found in Clemmens *et al.* [CLE 98]. This canal is a steep canal characterized by higher values of the Froude number than canal 2, although both canals operate in a subcritical flow. Canal 1 has little storage capacity due to its small cross-sections. On the other hand, canal 2 is based on the upstream portion of the Corning canal in California and is much flatter and has significant storage [BUY 79]. In both canals, the target depth is close to the normal depth of each pool at maximum flow.

10.1.1. *Maricopa Stanfield canal*

The Maricopa Stanfield canal is 9.5 km long. Rectangular gates divide it into eight pools. The orifice offtakes, emergency lateral spillways, pump stations and checkpoints or target points are located at the downstream end of each pool. The canal geometry is represented in Figure 10.1 and the canal features are shown in Tables 10.1 and 10.2.

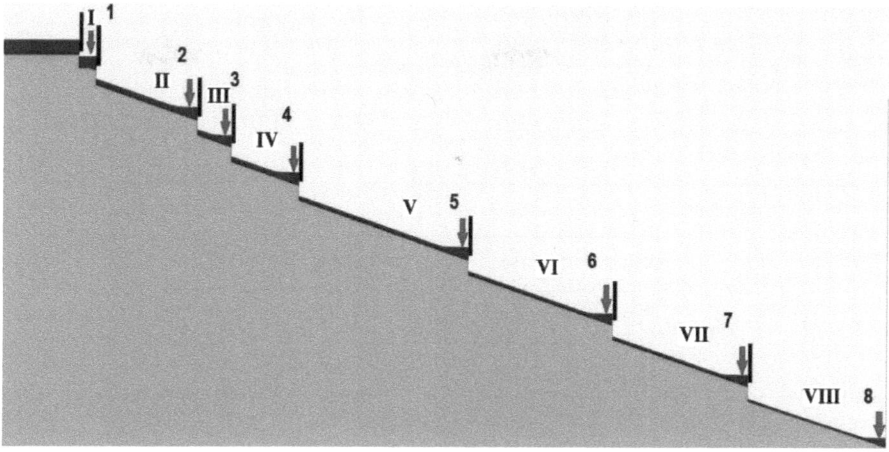

Figure 10.1. *Maricopa Stanfield in profile. The red lines mark the position of the checkpoints. The first pool is number I and the first checkpoint is number 1. For a color version of the figure, please see www.iste.co.uk/bonet/canals.zip*

Pool number	Numerical node upstream	Numerical node downstream	Pool length (km)	Bottom slope	Side slopes (H:V)	Manning's coefficient (n)	Bottom width (m)	Canal depth (m)
I	1	6	0.1	$2*10^{-3}$	1.5:1	0.014	1	1.1
II	7	67	1.2	$2*10^{-3}$	1.5:1	0.014	1	1.1
III	68	88	0.4	$2*10^{-3}$	1.5:1	0.014	1	1.0
IV	89	129	0.8	$2*10^{-3}$	1.5:1	0.014	0.8	1.1
V	130	230	2	$2*10^{-3}$	1.5:1	0.014	0.8	1.1
VI	231	316	1.7	$2*10^{-3}$	1.5:1	0.014	0.8	1.0
VII	317	397	1.6	$2*10^{-3}$	1.5:1	0.014	0.6	1.0
VIII	398	483	1.7	$2*10^{-3}$	1.5:1	0.014	0.6	1.0

Table 10.1. *Features of Maricopa Stanfield pools*

Target points	Numerical node upstream	Numerical node downstream	Gate discharge coefficient	Gate width (m)	Gate height (m)	Step (m)	Distance from gate 1 (km)	Orifice offtake height (m)
0	0	1	0.61	1.5	1.0	1.0	0	–
1	6	7	0.61	1.5	1.1	1.0	0.1	0.45
2	67	68	0.61	1.5	1.1	1.0	1.3	0.45
3	88	89	0.61	1.5	1.0	1.0	1.7	0.40
4	129	130	0.61	1.2	1.1	1.0	2.5	0.45
5	230	231	0.61	1.2	1.1	1.0	4.5	0.45
6	316	317	0.61	1.2	1.0	1.0	6.2	0.40
7	397	398	0.61	1.0	1.0	1.0	7.8	0.40
8	483	483	–	–	–	–	9.5	0.40

Table 10.2. *Maricopa Stanfield control structures*

10.1.1.1. *Discretization*

The prediction horizon is 12 hours (43200 seconds). The regulation time step is 300 seconds (five minutes); consequently, there are 144 regulation periods. There are eight sluice gates to control eight pools. As every gate position estimated during every regulation period, we need 1152 gate trajectory parameters (144 regulation periods by eight sluice gates). The canal is spatially discretized every 20 meters.

10.1.1.2. *Initial and boundary conditions*

Upstream of the first gate in Maricopa Stanfield canal, there is a reservoir and so the boundary condition is a constant water level of 1.0 m and this value remains constant over the irrigation cycle. The steady state is the initial condition at the beginning of the test cases; the initial condition is shown in Tables 10.3 (test case 1-1) and 10.4 (test case 1-2) and the downstream water level of every pool is defined by the target value.

Pool number	Offtake initial flow (m^3/s)	Resulting initial check flow (m^3/s)
Heading	–	1.0
1	0.1	0.9
2	0.1	0.8
3	0.2	0.6
4	0.2	0.4
5	0.1	0.3
6	0.1	0.2
7	0.1	0.1
8	0.1	0.0

Table 10.3. *Initial conditions of test case 1-1 (Maricopa Stanfield)*

Pool number	Offtake initial flow (m^3/s)	Resulting initial check flow (m^3/s)
Heading	–	2.0
1	0.2	1.8
2	0.0	1.8
3	0.4	1.4
4	0.0	1.4
5	0.0	1.4
6	0.3	1.1
7	0.2	0.9
8	0.9	0.0

Table 10.4. *Initial conditions of test case 1-2 (Maricopa Stanfield)*

10.1.2. *Corning canal*

The Corning canal of California has significant storage capacity. The canal's length is 28 km and the cross-sections are trapezoidal. Rectangular gates divide it into eight pools. There are orifice offtakes, emergency lateral spillways, pump stations and checkpoints or target points located at the downstream end of each pool. The canal geometry is shown in Figure 10.2 and some other features are shown in Tables 10.5 and 10.6.

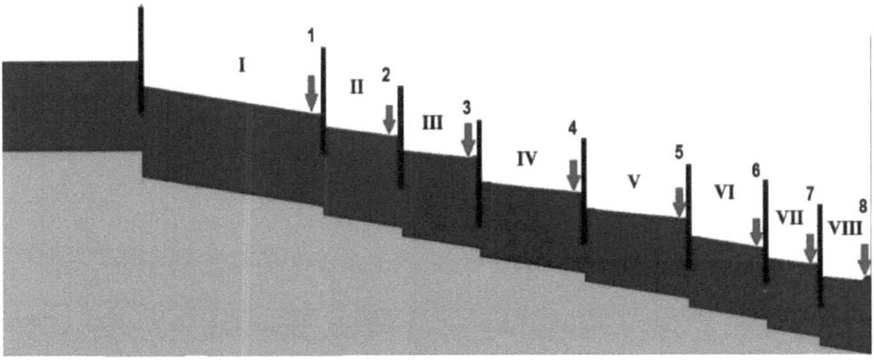

Figure 10.2. *Corning canal profile. The red lines mark the position of the checkpoints. The first pool is number I and the first checkpoint is number 1. For a color version of the figure, please see www.iste.co.uk/bonet/canals.zip*

Pool number	Numerical node upstream	Numerical node downstream	Pool length (km)	Bottom slope	Side slopes (H:V)	Manning's coefficient (n)	Bottom width (m)	Canal depth (m)
I	1	141	7	10^{-4}	1.5:1	0.02	7	2.5
II	142	202	3	10^{-4}	1.5:1	0.02	7	2.5
III	203	263	3	10^{-4}	1.5:1	0.02	7	2.5
IV	264	344	4	10^{-4}	1.5:1	0.02	6	2.3
V	345	425	4	10^{-4}	1.5:1	0.02	6	2.3
VI	426	486	3	10^{-4}	1.5:1	0.02	5	2.3
VII	487	527	2	10^{-4}	1.5:1	0.02	5	1.9
VIII	528	568	2	10^{-4}	1.5:1	0.02	5	1.9

Table 10.5. *Features of the Corning canal pools*

Target points	Numerical node upstream	Numerical node downstream	Gate discharge coefficient	Gate width (m)	Gate height (m)	Step (m)	Distance from gate 1 (km)	Orifice offtake height (m)	Lateral spillway height (m)
0	0	1	0.61	7	2.3	0.2	0	–	3
1	141	142	0.61	7	2.3	0.2	7	1.05	2.5
2	202	203	0.61	7	2.3	0.2	10	1.05	2.5
3	263	264	0.61	7	2.3	0.2	13	1.05	2.5
4	344	345	0.61	6	2.1	0.2	17	0.95	2.3
5	425	426	0.61	6	2.1	0.2	21	0.95	2.3
6	486	487	0.61	5	1.8	0.2	24	0.85	1.9
7	527	528	0.61	5	1.8	0.2	26	0.85	1.9
8	568	568	–	–	–	–	28	0.85	1.9

Table 10.6. *Corning canal control structures*

10.1.2.1. *Discretization*

The predictive horizon is 12 hours (43200 seconds). The regulation time step is 900 seconds (15 minutes) with 48 regulation periods. There are eight sluice gates to control the canal pools; 384 sluice gate trajectory parameters are needed (48 regulation periods by eight sluice gates). The canal is spatially discretized every 50 meters.

10.1.2.2. *Initial and boundary conditions*

Upstream of the first gate in the Corning canal, there is a reservoir, and so the boundary condition is a constant water level of 3.0 meters and this value remains constant over the predictive horizon. The steady state is the initial condition at the beginning of the test cases; the initial conditions are shown in Tables 10.7 (test 2-1) and 10.8 (test 2-2).

Pool number	Offtake initial flows (m^3/s)	Resulting initial check flows (m^3/s)
Heading	–	13.5
1	1.0	12.5
2	1.0	11.5
3	1.0	10.5
4	1.0	9.5
5	2.5	7.0
6	2.0	5.0
7	1.0	4.0
8	1.0	3.0

Table 10.7. *Initial conditions of test case 2-1 (Corning canal)*

Pool number	Offtake initial flows (m^3/s)	Resulting initial check flows (m^3/s)
Heading	–	13.7
1	1.7	12.0
2	1.8	10.2
3	2.7	7.5
4	0.3	7.2
5	0.2	7.0
6	0.8	6.2
7	1.2	5.0
8	0.3 +2.0[1]	2.7

Table 10.8. *Initial conditions of test case 2-2 (Corning canal)*

1 Changes in downstream pump discharge with no change in offtake flow.

The offtake flows are obtained by gravity offtakes located at the downstream end of each pool, where the extracted flow through the offtake ($Q_{offtake}$) is described as follows:

$$Q_{offtake} = K\sqrt{y - y_0}$$
$$K = Cd \times \frac{\pi \times \emptyset^2}{4} \times \sqrt{2g} \quad [10.1]$$

where y is the water level of the canal at the offtake; y_0 is the orifice offtake height equal to $y_{target}/2$; C_d is the discharge coefficient which is dependent of the Reynolds number and thus it has a variable value during the irrigation cycle; and Φ is the orifice offtake diameter.

10.2. Target values

The target values, which we have to maintain at the orifice offtake during the irrigation cycle, are fixed throughout the whole test. We introduce these values for every test case in this section.

10.2.1. *Maricopa Stanfield*

As stated above, the control algorithm has to maintain a desired water level at the checkpoints during a predictive horizon. The desired water levels are shown in Table 10.9.

Target point	Desired water level (m)
1	0.9
2	0.9
3	0.8
4	0.9
5	0.9
6	0.8
7	0.8
8	0.8

Table 10.9. *Desired water level for test case 1-1/1-2 (Maricopa Stanfield)*

10.2.2. *Corning canal*

The desired water levels at the checkpoints are shown in Table 10.10.

Target point	Water level desired (m)
1	2.1
2	2.1
3	2.1
4	1.9
5	1.9
6	1.7
7	1.7
8	1.7

Table 10.10. *Desired water level for test case 2-1/2-2 (Corning canal)*

10.3. Unscheduled flow changes

The duration of unscheduled deliveries is 10 hours, almost all of the test duration. It is not easy to control the canal because these disturbances cause hydraulic and numerical problems in the control algorithms.

The CSI algorithm calculates the unscheduled water deliveries and sends the output data to the control algorithm, which responds by modifying the gate trajectories. However, no one knows the total duration of the unscheduled water deliveries. For this reason, the GoRoSoBo algorithm calculates the sluice gate trajectories for a predictive horizon, taking into account the unscheduled water deliveries obtained by the CSI algorithm at the present time, and GoRoSoBo extrapolates the unscheduled water deliveries in all prediction horizons in order to calculate the sluice gate trajectories, as is described in Chapter 8. In the next regulation period, new water level measurements are taken at the checkpoints and the CSI algorithm comes back to calculate the new unscheduled water deliveries and the GoRoSoBo algorithm comes back to calculate the new gate trajectories as well. In the case where there are only unscheduled water deliveries at the present time and not during the following regulation periods, GoRoSoBo will modify the gate trajectories to take into account a wrong water level error in the prediction horizon. The errors caused by that hypothesis (extrapolate the unscheduled water deliveries to the whole prediction horizon) are small, although these errors depend on the duration of the regulation period and the value of the unscheduled water deliveries.

We had to consider this hypothesis because the sluice gate movements at the present time are essential to reducing disturbance in the future. The disturbance wave generated by sluice gate movement does not travel quickly on a free surface, so the control actions must be taken at the present time in order to obtain a response in the future. If we consider that the unknown delivery does not exist in the future, and that the unscheduled offtake changes are permanent, then the control algorithm can never obtain a gate trajectory to reduce the deviation between the desired and the measured water level.

This happens because once the disturbance disappears, the water level at the checkpoints quickly returns to the desired level by itself; if we make this supposition, then the control algorithm does not need to take action for the future. However, if this supposition is wrong and the disturbance lasts for a long time, then the water level error will not be removed. For this reason, as we explained in Chapter 8, the unscheduled offtake obtained by the CSI algorithm at the present time is extrapolated during the predictive horizon.

We must note that there are gates which have no influence on some target points at the end of the prediction horizon; for instance, a movement performed by the first gate in the last regulation period has no influence on target 8 in that regulation period because the wave generated by the first sluice gate in that period cannot reach target 8. There are some elements in the HIM matrix with zero values and these zeroes can produce numerical problems, even more so when significant unscheduled delivery is concerned, as there are long periods in the prediction horizon during which the sluice gate positions have changed notably from the initial gate position; whereas, at the end of the prediction horizon, as the sluice gates have no influence on the checkpoints, the sluice gates remain in their initial position. These changes in gate trajectory are significant enough to generate numerical problems. To solve these numerical problems, we increased the duration of the last regulation period four times, as this way all sluice gates can influence any checkpoint during the last regulation period.

10.3.1. *Maricopa Stanfield*

The unscheduled water deliveries are more important in test 1-2 than in test 1-1, as is shown in Tables 10.11 and 10.12. These unknown deliveries are introduced from the second hour of the test until the end of the test.

Pool number	Offtake initial flow (m^3/s)	Check initial flow (m^3/s)	Unscheduled offtake changes at 2 hours (m^3/s)	Check final flow (m^3/s)
Heading	–	1.0	–	0.8
1	0.1	0.9	–	0.7
2	0.1	0.8	–	0.6
3	0.2	0.6	–	0.4
4	0.2	0.4	–0.1	0.3
5	0.1	0.3	–0.1	0.3
6	0.1	0.2	–	0.2
7	0.1	0.1	–	0.1
8	0.1	0.0	–	0.0

Table 10.11. *Unscheduled offtake changes in test case 1-1 (Maricopa Stanfield)*

Pool number	Offtake initial flow (m^3/s)	Check initial flow (m^3/s)	Unscheduled offtake changes at 2 hours (m^3/s)	Check final flow (m^3/s)
Heading	–	2.0	–	2.0
1	0.2	1.8	–	1.8
2	0.0	1.8	0.2	1.6
3	0.4	1.4	–0.2	1.4
4	0.0	1.4	0.2	1.2
5	0.0	1.4	0.2	1.0
6	0.3	1.1	–0.1	0.8
7	0.2	0.9	–	0.6
8	0.9	0.0	–0.3	0.0

Table 10.12. *Unscheduled offtake changes in test case 1-2 (Maricopa Stanfield)*

The unscheduled water deliveries in test 1-2 are important despite the changes in the total flow rate being null, because there are pools where the unscheduled deliveries are positive whilst others are negative.

The unscheduled deliveries are more significant at target 8 in test case 1-2, where the flow rate changes from 0.9 m^3/s to 0.6 m^3/s (33%). For this reason, the control algorithm has to quickly modify the gate trajectories, because there is the possibility that the canal will overflow into some pools downstream due to the low storage capacity. This is more common in canals like Maricopa Stanfield rather than the Corning canal, which has a significant amount of storage capacity.

10.3.2. *Corning canal*

In test 2-1, we find that the unscheduled water deliveries are not very significant, as is shown in Tables 10.13 and 10.14. The only change in flow is in the orifice offtake of pool six. This is a change in the total flow in the canal of 14.8% and in the total flow in pool six of 28.5%. These unknown deliveries are introduced from the second hour of the test until the end. These unscheduled deliveries are flow rate changes, and the control algorithm regards or considers these flow changes once the water levels increase or decrease at target points or checkpoints.

Pool number	Offtake initial flow (m^3/s)	Check initial flow (m^3/s)	Unscheduled offtake changes at 2 hours (m^3/s)	Resulting check flow (m^3/s)
Heading	–	13.5	–	11.5
1	1.0	12.5	–	10.5
2	1.0	11.5	–	9.5
3	1.0	10.5	–	8.5
4	1.0	9.5	–	7.5
5	2.5	7.0	–	5.0
6	2.0	5.0	–2	5.0
7	1.0	4.0	–	4.0
8	1.0	3.0	–	3.0

Table 10.13. *Unscheduled offtake changes on test case 2-1 (Corning canal)*

Pool number	Offtake initial flow (m^3/s)	Check initial flow (m^3/s)	Unscheduled offtake changes at 2 hours (m^3/s)	Resulting check flow (m^3/s)
Heading	–	13.7	–	2.7
1	1.7	12.0	– 1.5	2.5
2	1.8	10.2	– 1.5	2.2
3	2.7	7.5	– 2.5	2.0
4	0.3	7.2	–	1.7
5	0.2	7.0	–	1.5
6	0.8	6.2	– 0.5	1.2
7	1.2	5.0	– 1.0	1.0
8	0.3 +2.0*	2.7	– 2.0[2]	0.7

Table 10.14. *Unscheduled offtake changes on test case 2-2 (Corning canal)*

The unscheduled water deliveries in test 2-2 are very relevant, as these water deliveries make up a significant percentage of the total flow. In this case, the unscheduled water deliveries make up 81% of the total flow and these water changes are more significant at target 8, where the flow rate changes from 5 m^3/s to 1 m^3/s. For this reason, it should be pointed out that 15 minutes after introducing the unscheduled water deliveries into the canal, the water level in target 8 is 1.93 m, so the lateral spillway must operate. This is the maximum change in the flow rate in all test cases.

10.4. Constraints and trajectories

According to Soler [SOL 03], there are different ways to reach the final solution, depending on the maximum or minimum gate position, the maximum sluice gate movement permitted between consecutive regulation periods, or the time-frequency at which the actuator can move the gates. The larger the maximum sluice gate movement permitted, the quicker it is to drive the state of the canal toward the

2 Changes in downstream pump discharge with no change in offtake flow.

desired state in the case of significant flow changes; on the other hand, however, it is easier to damage the servomechanism of the sluice gate because the gate trajectories oscillate more. In such cases, an optimization problem in which the constraints are too lax can destabilize the model, because the change of the free surface caused by the gate trajectories can be very abrupt.

The ASCE restricts the minimum regulation period to 15 minutes for test case 2-1/2-2. This was the regulation period chosen in tests 2-1 and 2-2, and the time-frequency on gate movement is also 15 min. The ASCE restricts the minimum regulation period to 5 minutes for tests 1-1 and 1-2, and so this was the regulation period chosen in tests 1-1 and 1-2 and the time-frequency on sluice gate movement.

Considering the results obtained in the analysis of constraints carried out by Soler [SOL 03] and Soler *et al.* [SOL 13], we have imposed the overall and functional constraints in Table 10.15. The constraints are determined as a certain percentage of the gate height and these constraints are imposed on the gate positions.

	U_{min} (%)	U_{max} (%)	dU_{max}/ $dU0_{max}$ (%)
Maricopa Stanfield	2	90	5
Corning canal	0.5	90	5

Table 10.15. *Overall and functional constraint values*

The value of the minimum gate position is different in both cases. The gate width of the Corning canal is almost five times larger than that of the Maricopa Stanfield, and so the minimum gate position is fixed by a combination of the gate width and the flow through the gate. For instance, we cannot fix a minimum gate position of 2% for the Corning canal, because this gate position is equivalent to a significant flow through the gate; in situations where the total flow in the canal decreases by an important reduction in water demand, the gate's position should be lower than 2%.

10.5. Test results

We show the results in three graphs for every test case:

– The first graph shows the scheduled and unscheduled offtake changes (demanded extracted flow), the real value delivered by the offtake after introducing the new gate trajectories by GoRoSoBo during the irrigation cycle, and the real extracted flow calculated by CSI. As we introduced before, GoRoSoBo needs the CSI information in order to calculate the gate trajectories; therefore, CSI has to calculate the extracted flow in the canal, taking the scheduled and unscheduled offtake flows into consideration.

– The second graph[3] shows the gate trajectories during the irrigation cycle, where the gate trajectory is defined by the gate position in time, and the gate position is defined as follows:

$$Gate\ position = \frac{Gate\ opening}{Gate\ height} \qquad [10.2]$$

– The third graph shows the water level at the checkpoints, which is proportional to the real flow rate delivered to the farmers by the offtake.

10.5.1. *Test case 1-1 (Maricopa Stanfield)*

In Table 10.16, we show the scheduled offtakes and the unscheduled water changes which are introduced two hours after starting the test.

Pool number	Scheduled offtake flow (m^3/s)	Unscheduled offtake changes at 2 hours (m^3/s)
Heading	–	–
1	0.1	–
2	0.1	–
3	0.2	–
4	0.2	– 0.1
5	0.1	– 0.1
6	0.1	–
7	0.1	–
8	0.1	–

Table 10.16. *Unscheduled offtake changes in test case 1-1 (Maricopa Stanfield)*

3 Please note that the gate trajectories and water level trajectories have been divided into two graphs, the second and third, for better comprehension of the results.

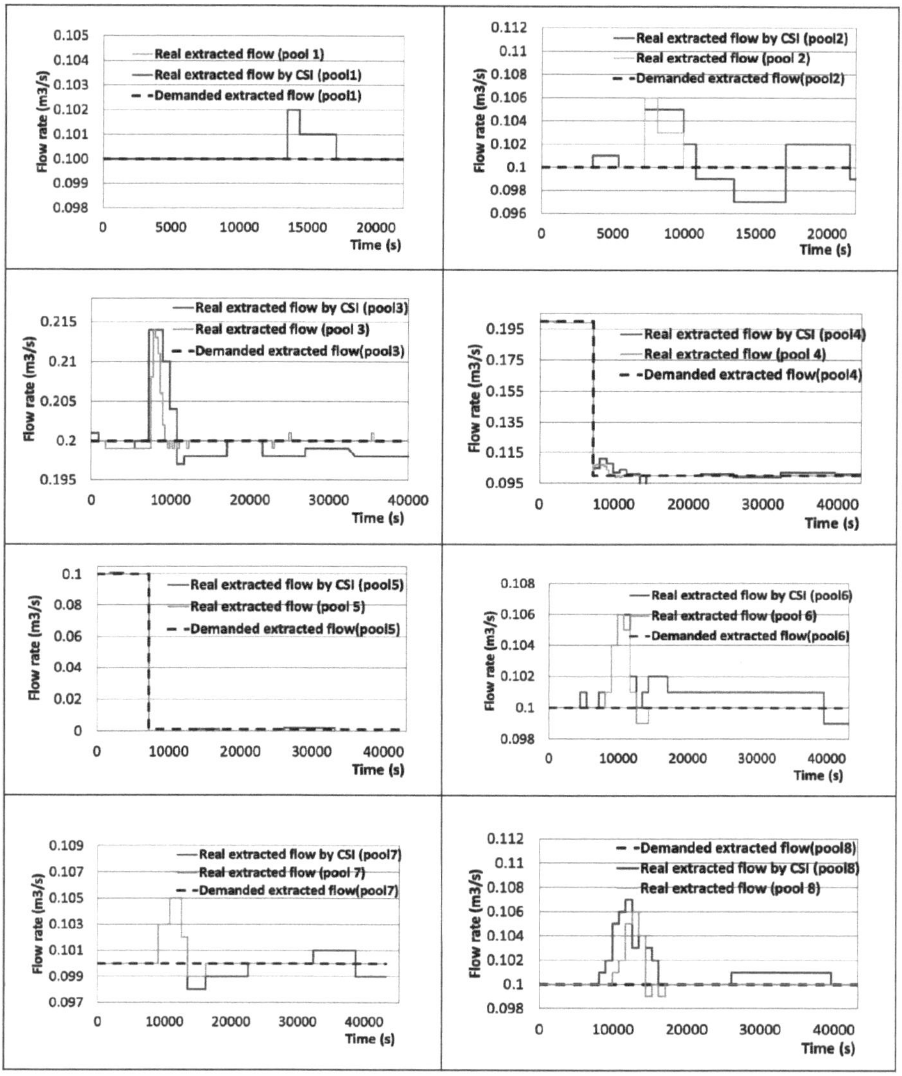

Figure 10.3. *Demanding extracted flow, extracted flow by CSI and real extracted flow in test case 1-1 (Maricopa Stanfield). For a color version of the figure, please see www.iste.co.uk/bonet/canals.zip*

The flow changes are only introduced in the fourth and fifth pools (Table 10.16), and CSI is able to perfectly establish the extracted flow in the canal with a maximum numerical error of 4% between the demanded extracted flow and the value produced by CSI; see Figure 10.3. This small error is due to numerical errors in CSI, and we obtain a similar maximum error in every pump station.

Once the GoRoSoBo algorithm realizes that the water level increases at checkpoints 4 and 5 at 7500 s (Figures 10.4 and 10.5), sluice gates 1 to 5 start to close in order to reduce the water level at the checkpoints downstream (Figures 10.6 and 10.7), and sluice gates 6, 7 and 8 start to open to reduce the water level at the checkpoints upstream. After several regulation periods, sluice gates 6, 7 and 8 must close because the water level at checkpoints 6, 7 and 8 also increase, see Figure 10.6 and 10.7 (10.000 seconds).

GoRoSoBo starts to modify the gate trajectories at 7500 s; one regulation period after the disturbance is introduced into the canal. GoRoSoBo drives the canal to the desired water level at every checkpoint at 18000 s.

At the end, the controller response is smoother and gate movement is damped to prevent oscillations that would cause excessive control action, because the water levels are almost equal to the desired levels with a maximum error around 0.26%.

The results obtained in this test case are quite good, as the performance indicators demonstrate; see section 10.6. Four hours later, when the unscheduled offtake changes are introduced, GoRoSoBo drives the canal to the target value.

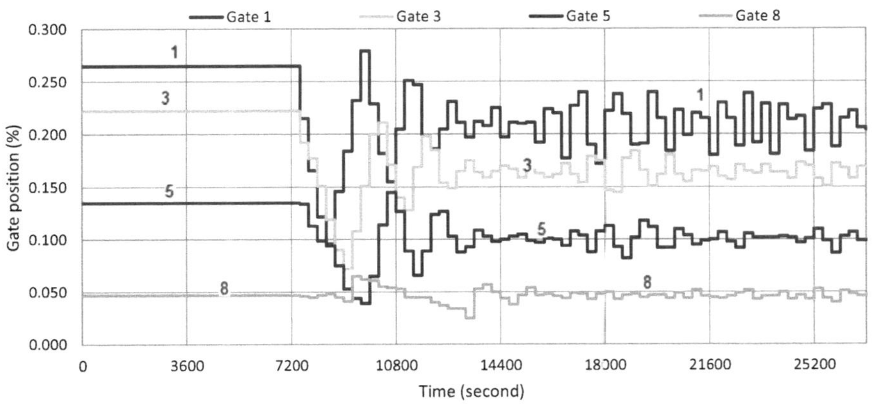

Figure 10.4. *Gate trajectories (1, 3, 5 and 8) in test case 1-1 (Maricopa Stanfield). For a color version of the figure, please see www.iste.co.uk/bonet/canals.zip*

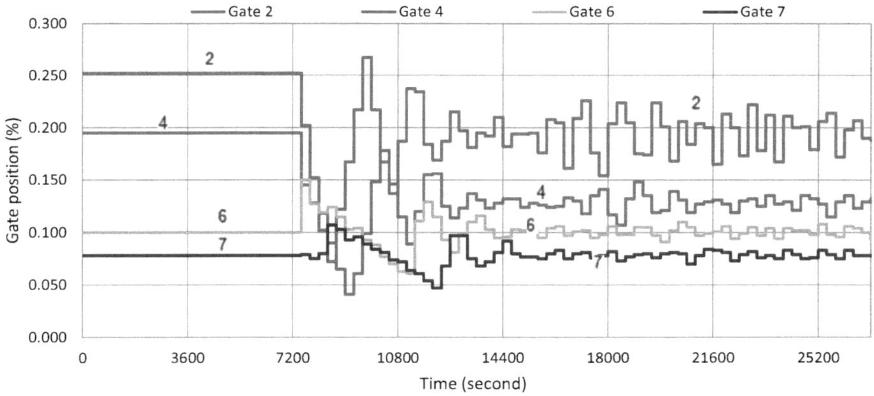

Figure 10.5. *Gate trajectories (2, 4, 6 and 7) in test case 1-1 (Maricopa Stanfield). For a color version of the figure, please see www.iste.co.uk/bonet/canals.zip*

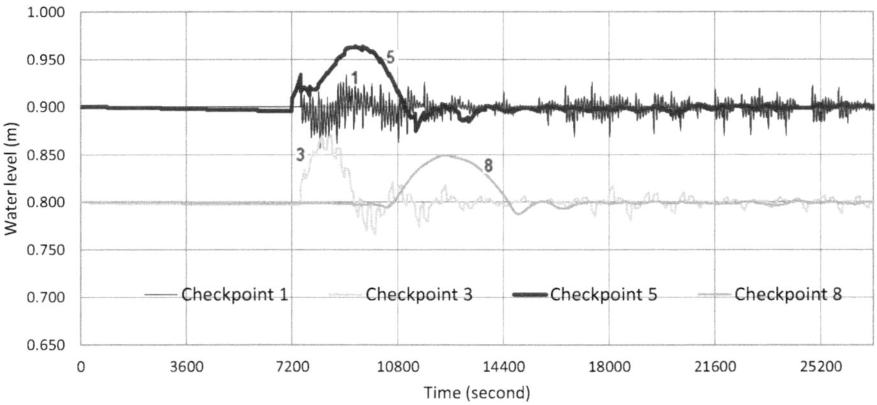

Figure 10.6. *Water level at checkpoints 1, 3, 5 and 8 in test case 1-1 (Maricopa Stanfield). For a color version of the figure, please see www.iste.co.uk/bonet/canals.zip*

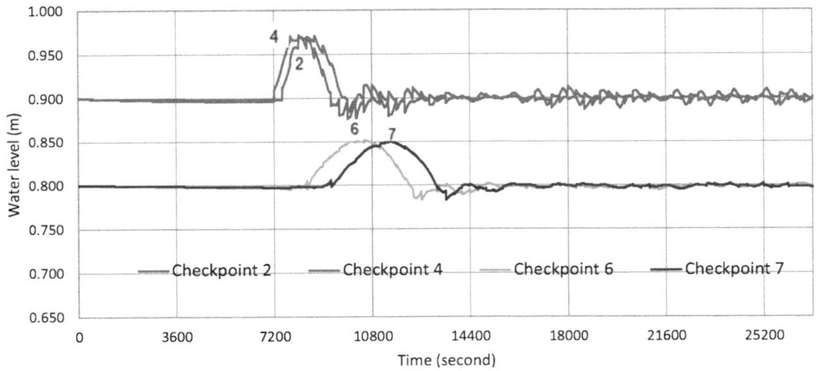

Figure 10.7. *Water level at checkpoints 2, 4, 6 and 7 in test case 1-1 (Maricopa Stanfield). For a color version of the figure, please see www.iste.co.uk/bonet/canals.zip*

10.5.2. *Test case 1-2 (Maricopa Stanfield)*

The scheduled offtakes and the unscheduled water changes, which are introduced two hours after starting the test, are shown in Table 10.17.

Pool number	Scheduled offtake changes (m^3/s)	Unscheduled offtake changes at 2 hours (m^3/s)
Heading	–	–
1	0.2	–
2	0.0	0.2
3	0.4	– 0.2
4	0.0	0.2
5	0.0	0.2
6	0.3	– 0.1
7	0.2	–
8	0.9	– 0.3

Table 10.17. *Unscheduled offtake changes in test case 1-2 (Maricopa Stanfield)*

Test 1-2 is much more difficult than test 1-1, because the unscheduled water deliveries are higher at all targets. The canal is in steady state during the first two hours. After the first two hours, unscheduled water deliveries are introduced to the system at 7200 s, although the algorithm is not notified until the next regulated period, once the water level is measured at the checkpoints. It is important to remark that the unscheduled water deliveries are relevant at all targets but especially at target 8, because in just one regulation period, the water level at checkpoint 8 increases from 0.8 m to 1.08 m (Figure 10.11).

A water delivery change of 0.3 m^3/s is introduced at checkpoint 8, with a total flow rate change at target 8 from 0.6 m^3/s to 0.9 m^3/s, that is, a flow rate change of around 33.3%. The extracted flow is calculated by CSI, which establishes the value with a maximum numerical error of 2% between the demanded extracted flow and the value produced by CSI, as shown in Figure 10.8.

The unscheduled offtake changes are 0.2 m^3/s at targets 1 to 5, 0.1 m^3/s at target 6 and 0.3 m^3/s at target 8, in a canal with a total flow rate of 2 m^3/s which has no storage capacity and a steep slope; this is a difficult test for the control algorithm.

Once the GoRoSoBo algorithm knows that the water level is increasing less at checkpoints 2, 4 and 5 at time 7500 s (Figures 10.11 and 10.12), all gates below 7 and 8 have a tendency to close (Figures 10.9 and 10.10). Although the water level is increasing at several checkpoints, the maximum water level variation is at checkpoint 8 at 8100 s; for that reason, gates 7 and 8 close, whereas gates 1, 2, 3, 4, 5 and 6 open to readjust the water level at all checkpoints from 8100 s to 9900 s. We can check that the lower gate position at 18000 s is gate 8, as is logical, because the more significant flow change is at checkpoint 8.

The sluice gate trajectories have the same shape as the water level at the checkpoints, due to the gate trajectories and the water levels fluctuating around the desired solution.

The water level decreases at all targets at 9900 s due to changes in gate trajectories, and the water level returns to the desired level at all checkpoints at 21600 s.

The water levels were almost equal to the desired levels at the end of the test, with a maximum error of around 0.25%. The results obtained in this test case, 1-2, are as good as the results obtained in test case 1-1. GoRoSoBo needs one more hour to drive the system to the target value than in test case 1-1, but the unscheduled offtake changes are more important in test case 1-2.

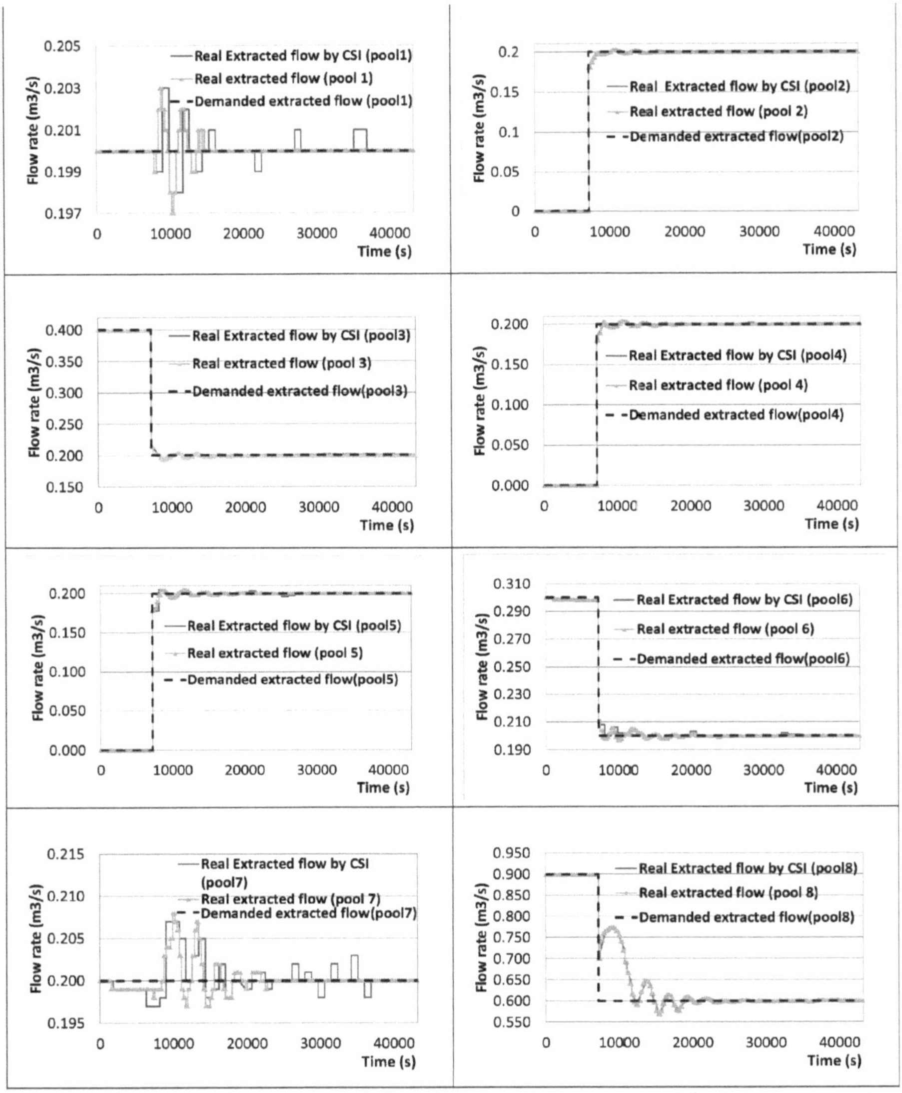

Figure 10.8. *Demanded extracted flow, extracted flow calculated by CSI and real extracted flow in test case 1-2 (Maricopa Stanfield). For a color version of the figure, please see www.iste.co.uk/bonet/canals.zip*

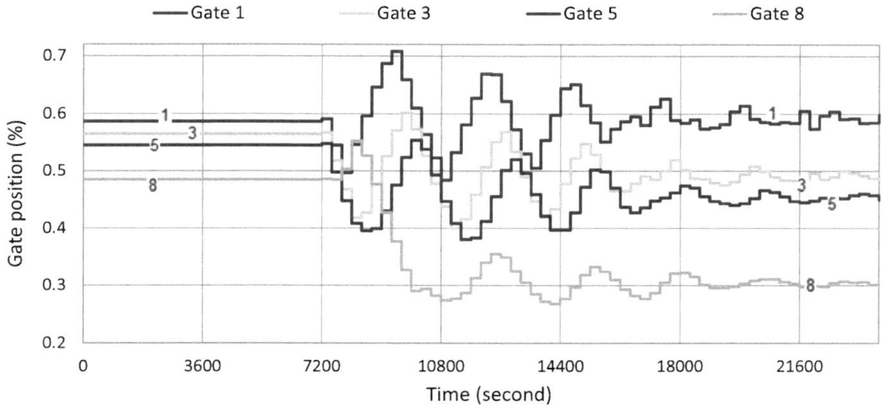

Figure 10.9. *Gate trajectories (1, 3, 5 and 8) in test case 1-2 (Maricopa Stanfield). For a color version of the figure, please see www.iste.co.uk/bonet/canals.zip*

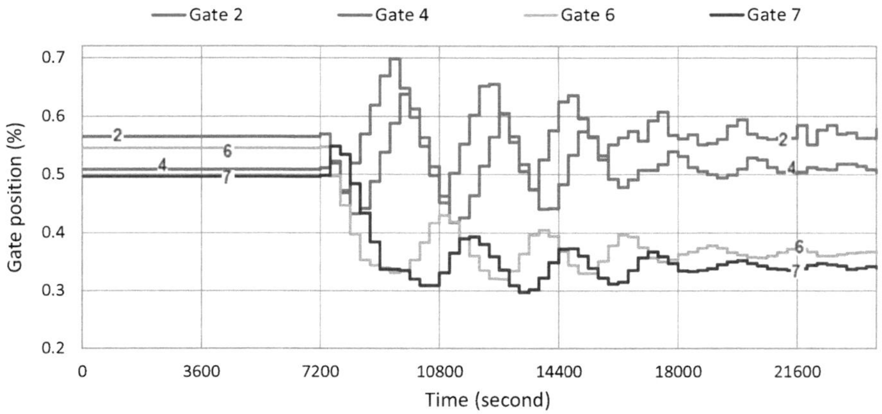

Figure 10.10. *Gate trajectories (2, 4, 6 and 7) in test case 1-2 (Maricopa Stanfield). For a color version of the figure, please see www.iste.co.uk/bonet/canals.zip*

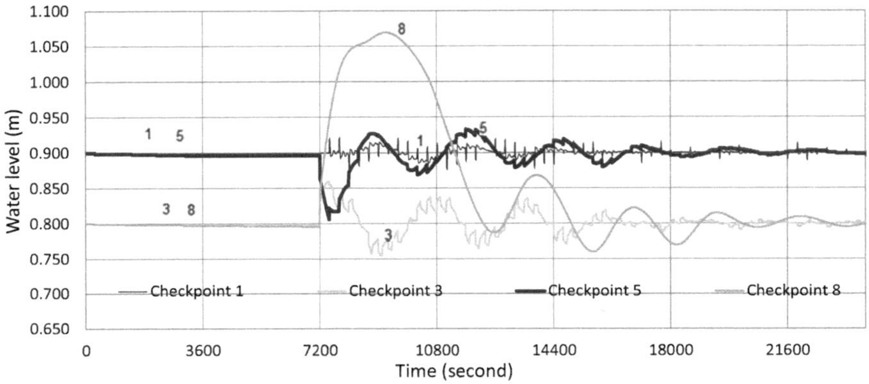

Figure 10.11. *Water level at checkpoints 1, 3, 5 and 8 in test case 1-2 (Maricopa Stanfield). For a color version of the figure, please see www.iste.co.uk/bonet/canals.zip*

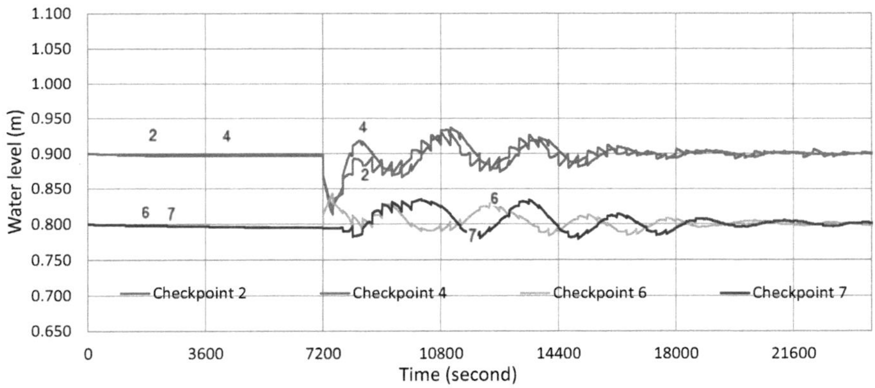

Figure 10.12. *Water level at checkpoints 2, 4, 6 and 7 in test case 1-2 (Maricopa Stanfield). For a color version of the figure, please see www.iste.co.uk/bonet/canals.zip*

10.5.3. *Test case 2-1 (Corning canal)*

The scheduled offtakes and the unscheduled water changes, which are introduced two hours after starting the test, are shown in Table 10.18.

Pool number	Scheduled offtake changes (m^3/s)	Unscheduled offtake changes at 2 hours (m^3/s)
Heading	–	–
1	1.0	–
2	1.0	–
3	1.0	–
4	1.0	–
5	2.5	–
6	2.0	– 2
7	1.0	–
8	1.0	–

Table 10.18. *Unscheduled offtake changes in test case 2-1 (Corning canal)*

The initial condition is the steady state. The sluice gates remain in their positions during the first two hours. The unscheduled water deliveries are introduced into the system at time 7200 s, although the algorithm is not notified until the next regulation period, when the water level at checkpoint 6 has already increased close to 10 cm. The measurements are sent to CSI, which establishes the extracted flow with a maximum numerical error of 0.7% between the demanded extracted flow and the value produced by CSI, as shown in Figure 10.13.

The change in water delivery is quite significant at target 6, as a water delivery change of 2 m^3/s is introduced and the total flow rate at target 6 is 7 m^3/s (that is, a flow rate change of 28% of the total flow rate).

Once the GoRoSoBo algorithm realizes that the water level has increased at checkpoint 6 at 8100 s (Figure 10.16), gates 1 to 6 start to close (Figures 10.14 and 10.15), and gates 7 and 8 start to open from 8100 s to 9900 s to reduce the increased water level at checkpoint 6 (Figure 10.14), and so the water level at checkpoints 7 and 8 starts to increase at 9000 s (Figure 10.17) and gates 7 and 8 start to close at 9900 s. The water level starts to decrease at all checkpoints at 12600 s, and the water level returns to the desired level at 19800 s; see Figures 10.16 and 10.17. The gate movements were almost zero in the last six hours of the test because the water levels at the checkpoints were equal to the desired water levels, with a maximum error of 0.2%. These small errors are due to residual errors in the computation of both algorithms (CSI and GoRoSoBo).

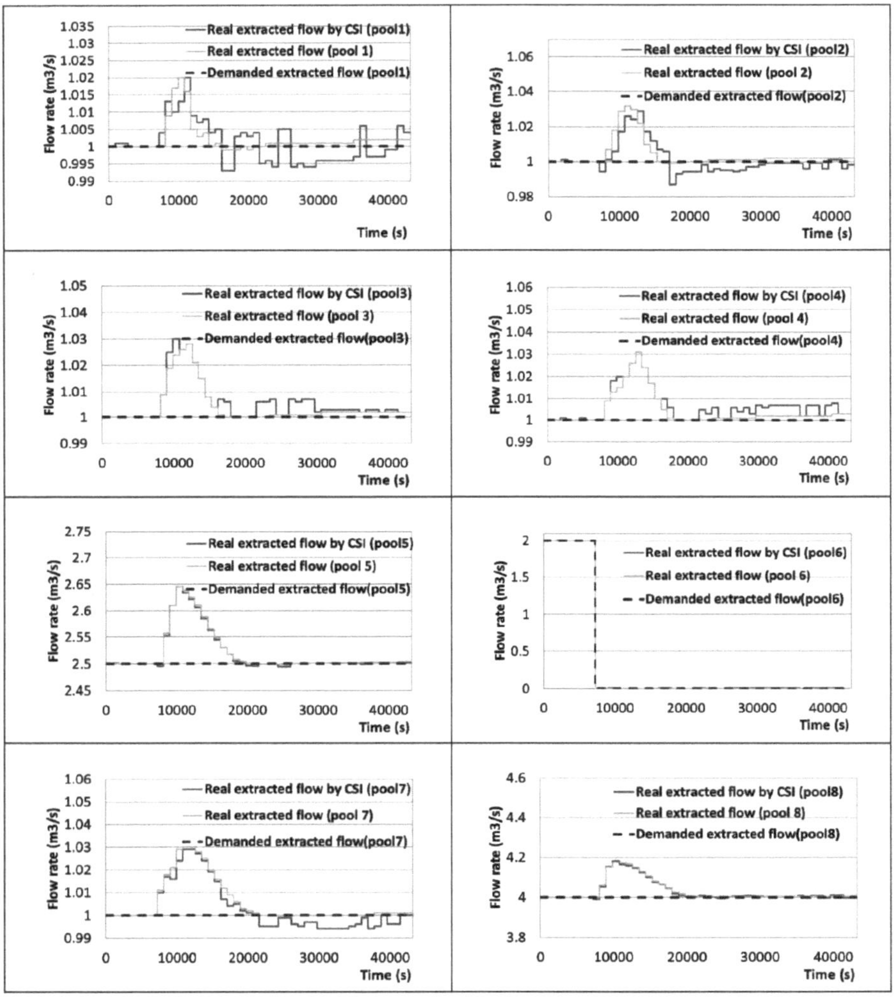

Figure 10.13. *Demanded extracted flow, extracted flow calculated by CSI and the real extracted flow in test case 2-1 (Corning canal). For a color version of the figure, please see www.iste.co.uk/bonet/canals.zip*

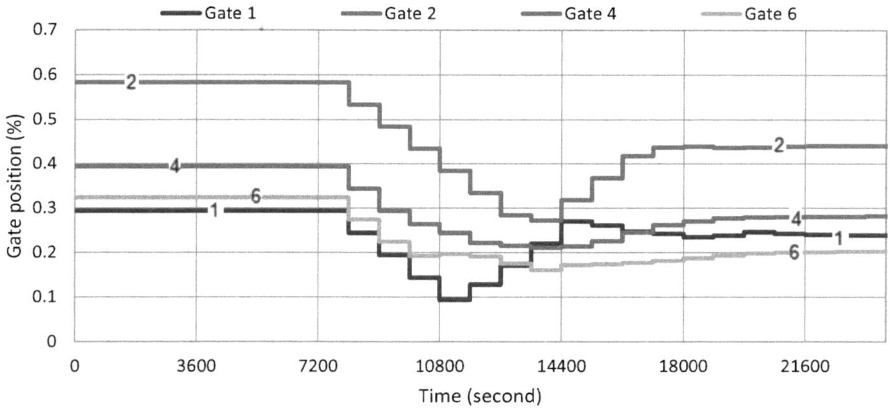

Figure 10.14. *Gate trajectories (1, 2, 4 and 6) in test case 2-1 (Corning canal). For a color version of the figure, please see www.iste.co.uk/bonet/canals.zip*

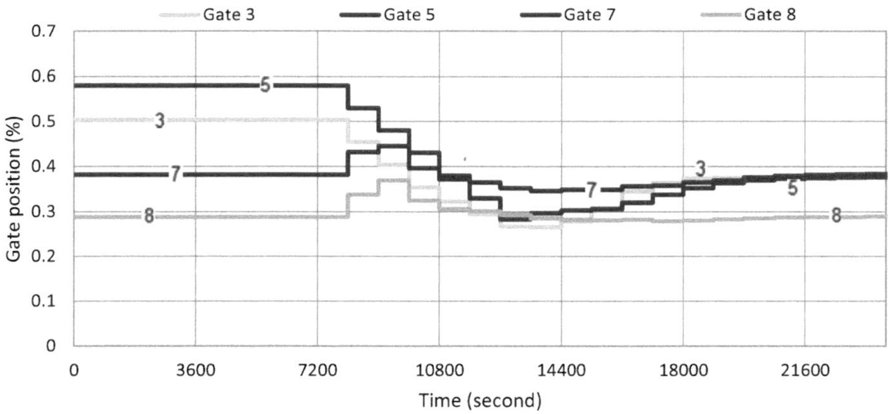

Figure 10.15. *Gate trajectories (3, 5, 7 and 8) in test case 2-1 (Corning canal). For a color version of the figure, please see www.iste.co.uk/bonet/canals.zip*

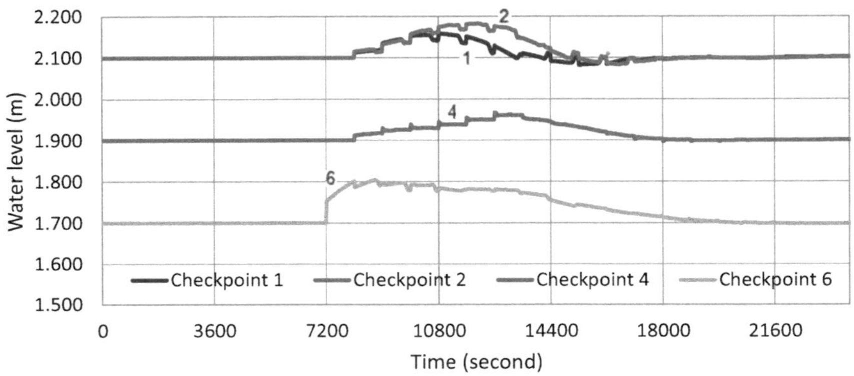

Figure 10.16. *Water level at checkpoints 1, 2, 4 and 6 in test case 2-1 (Corning canal). For a color version of the figure, please see www.iste.co.uk/bonet/canals.zip*

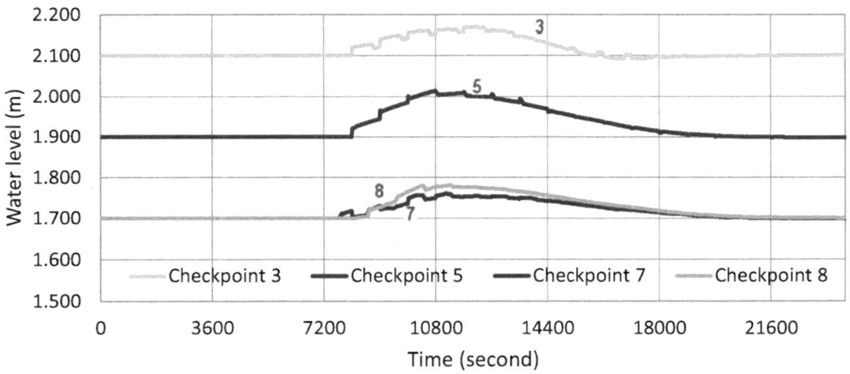

Figure 10.17. *Water level at checkpoints 3, 5, 7 and 8 in test case 2-1 (Corning canal). For a color version of the figure, please see www.iste.co.uk/bonet/canals.zip*

We can easily check the good behavior of CIS and GoRoSoBo in test case 2-1 because there is only a flow change at checkpoint 6 and the algorithm modifies all gates according to this one flow change.

10.5.4. *Test case 2-2 (Corning canal)*

In Table 10.19, we show the scheduled offtakes and the unscheduled water changes which are introduced two hours after starting the test.

Pool number	Scheduled offtake changes (m^3/s)	Unscheduled offtake changes at 2 hours (m^3/s)
Heading	–	–
1	1.7	–1.5
2	1.8	–1.5
3	2.7	–2.5
4	0.3	–
5	0.2	–
6	0.8	–0.5
7	1.2	–1.0
8	0.3 +2.0*	–2.0*

*Changes in downstream pump discharge with no change in offtake flow. The outlet flow at checkpoint 8 at initial time step is 2.7 m^3/s and the outlet flow at checkpoint 8 at 7200 s is 0.7 m^3/s.

Table 10.19. *Unscheduled offtake changes in test case 2-2 (Corning canal)*

Test 2-2 is the most difficult test due to the significant number of unscheduled water deliveries at all targets. The canal state is steady during the first two hours, as in the previous test cases. There are no unscheduled water deliveries for the first two hours, and so the sluice gates remain in their initial positions. After the first two hours, unscheduled water deliveries are introduced to the system. The water level measurements are sent to CSI, which establishes the real extracted flow with a maximum error of 3% from the demanded extracted flow, as shown in Figure 10.18. The unscheduled water deliveries are so significant at all targets that the total unscheduled deliveries were 9 m^3/s in a canal with a total flow of 13.7 m^3/s. GoRoSoBo obtained good results in test case 2-2, as shown in Figures 10.21 and 10.22, although it is important to remark that the unscheduled water deliveries are so significant, especially at target 8, that once the algorithm is notified of the rising water level at checkpoint 8 at 8100 s, the water level has already increased by 20 cm, that is, it reaches the top of the cross-section. In those circumstances, GoRoSoBo closes sluice gates 1 to 3 and opens sluice gates 4 to 8 (Figures 10.19 and 10.20), and thus in this way, the sluice gate trajectories reduce the increased water level at all checkpoints.

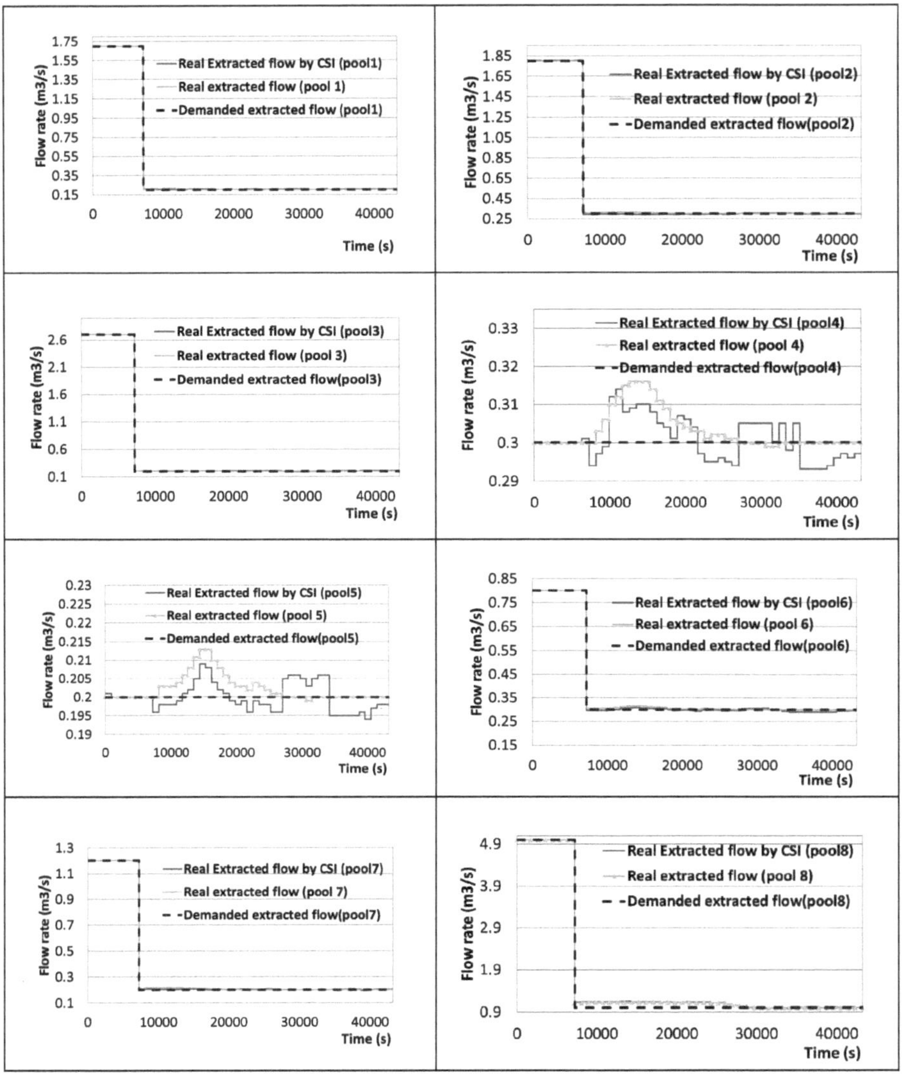

Figure 10.18. *Demanded extracted flow, extracted flow calculated by CSI and the real extracted flow in test case 2-2 (Corning canal). For a color version of the figure, please see www.iste.co.uk/bonet/canals.zip*

Redirecting the water level at checkpoints to the desired water level in a centralized system is not the duty of only one sluice gate. It is possible that some sluice gates have increased the water level error at some checkpoints during a regulation period (for instance, at checkpoints 4 and 5) when there are no unscheduled offtake changes at these checkpoints; however, the main objective is to redirect the measured water level to the desired values at all checkpoints as soon as possible.

The water level decreases at all checkpoints at 25200 s and the water level returns to the desired value at all of the checkpoints at 30600 s. The unscheduled water delivery is quite significant at target 8, introducing a total flow change of 4 m^3/s at 7200 s, because the total flow rate at pool 8 is 5 m^3/s at the initial time step and the flow rate at pool 8 is 1 m^3/s at 7200 s, and so the flow change is close to 80% of the total flow rate of the pool.

The gate movements are almost zero in the last two hours of the test because the water levels at the checkpoints are equal to the desired values at the end of the test, with a maximum error of around 0.23%.

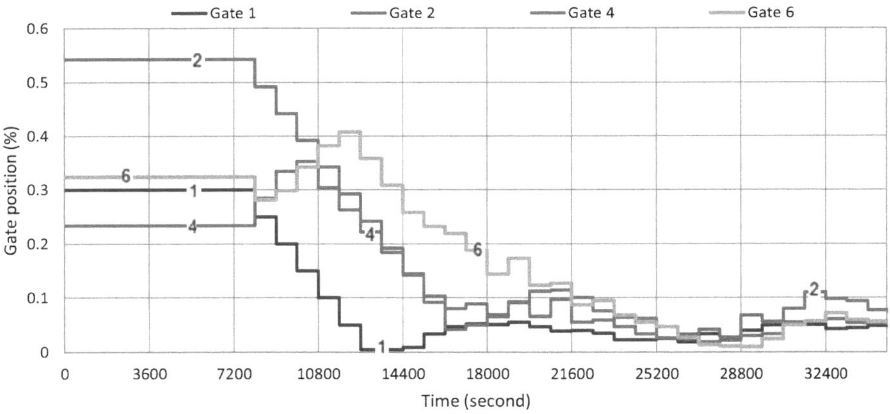

Figure 10.19. *Gate trajectories (1, 2, 4 and 6) in test case 2-2 (Corning canal). For a color version of the figure, please see www.iste.co.uk/bonet/canals.zip*

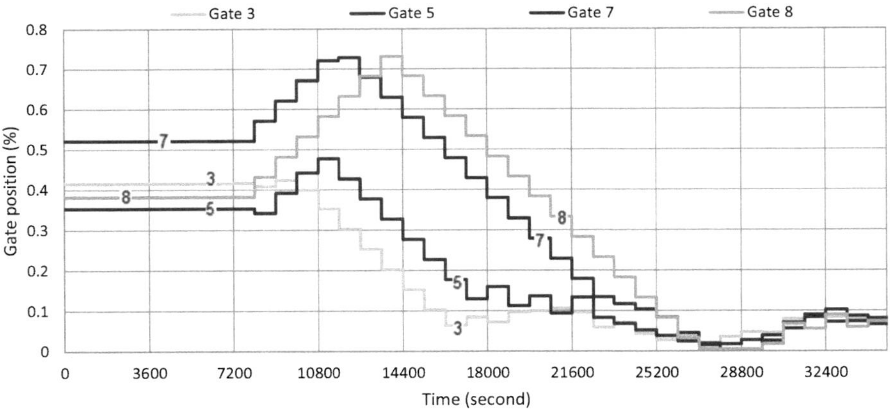

Figure 10.20. *Gate trajectories (3, 5, 7 and 8) in test case 2-2 (Corning canal). For a color version of the figure, please see www.iste.co.uk/bonet/canals.zip*

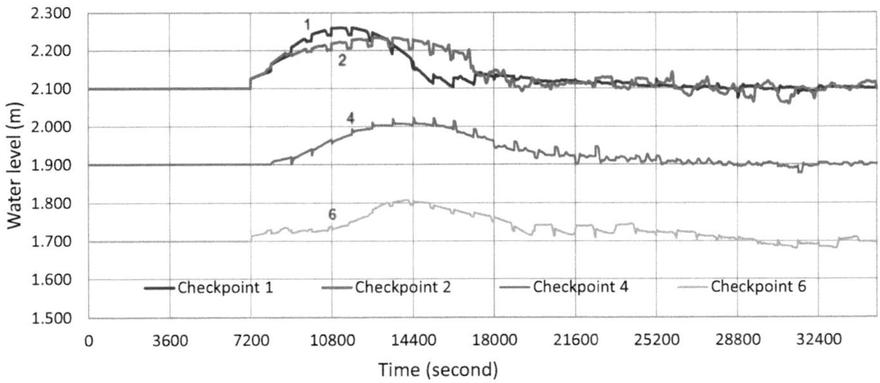

Figure 10.21. *Water level at checkpoints 1, 2, 4 and 6 in test case 2-2 (Corning canal). For a color version of the figure, please see www.iste.co.uk/bonet/canals.zip*

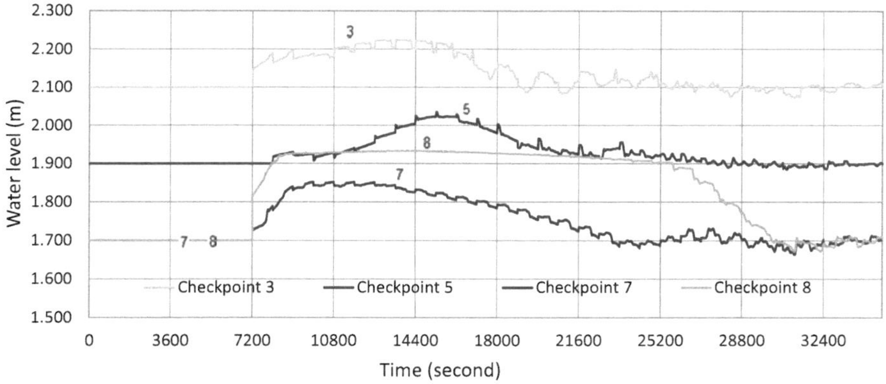

Figure 10.22. *Water level at checkpoints 3, 5, 7 and 8 in test case 2-2 (Corning canal). For a color version of the figure, please see www.iste.co.uk/bonet/canals.zip*

10.6. Performance indicators

These indicators were introduced to compare different control algorithms, as it is not easy to evaluate them and judge the controller's ability to deliver water. This is why the ASCE Task Committee devised these test cases.

Although not all of these performance indicators evaluate the deviations between the measured and the desired water level at the checkpoints, they also take into account other variables such as changes in flow.

There are many ways in which to assess the error between the measured and the desired water level at the checkpoints. First, we might calculate the maximum error at the checkpoint. Second, we might calculate how much time the control algorithm requires in order to remove the error at the checkpoint. Third, we might calculate the water level error at the checkpoint at the end of the test, since the water level should not stabilize at a wrong level. The fourth performance indicator controls the variations in the flow rate between the initial flow through the gate and the final flow. These four performance indicators are recommended by the ASCE Task Committee, and we show the maximum and the average of these performance indicators, as suggested by the Task Committee.

10.6.1. *Maximum absolute error (MAE)*

$$MAE = \frac{\max(|y_t - y_{target}|)}{y_{target}} \quad [10.3]$$

where y_t is the observed or measured (computed from simulation) water level at time t, and y_{target} is the target or the desired water level.

10.6.2. *Integral of absolute magnitude of error (IAE)*

$$IAE = \frac{\frac{\Delta t}{T}\sum_{t=0}^{T}|y_t - y_{target}|}{y_{target}} \quad [10.4]$$

where Δt is the regulation time step, and T is the time period for the test (12 hours). The time step must be held constant over the 12-hour interval specified.

10.6.3. *Steady-state error (StE)*

The steady-state error is defined as the maximum of the average error over the last two hours of the test. The conclusion is that the conditions should be stable during this period of time:

$$StE = \frac{\max(|y_{22,24} - y_{target}|)}{y_{target}} \quad [10.5]$$

where $y_{22,24}$ is the average water level between 10 and 12 hours. Some control algorithms have not performed well due to constant oscillation around the sluice gate trajectory solution. This indicator is proposed in order to detect excessive oscillations in water levels at the end of the test.

10.6.4. *Integrated absolute discharge change (IAQ)*

To avoid excessive oscillation of sluice gates during the irrigation cycle, this performance indicator, related to the changes in discharge due to sluice gate movement, is proposed:

$$IAQ = \sum_{t=t_1}^{t_2}(|Q_t - Q_{t-1}|) - |Q_{t_1} - Q_{t_2}| \quad [10.6]$$

where Q_t is the check gate discharge at time t, and so Q_{t-1} is the gate discharge at the previous time step. The second term is simply the difference between the initial and the final flow rate for the irrigation cycle. In that sense, the performance indicator

IAQ could be zero, both in the case where $Q_{t1} > Q_{t2}$ and $Q_{t-1} > Q_t$ for all time steps during the period of time (t_1 - t_2) and in the case where $Q_{t1} < Q_{t2}$ and $Q_{t-1} < Q_t$ for all time steps during that same period of time. Use of this performance indicator requires that the times for flows introduced in the summation include the times at all reversals in the sign of flow changes (all of the peaks and valleys of the flow hydrograph are included to set this performance indicator).

10.7. Comparison between controllers

We compare the performance indicators obtained with GoRoSoBo in test cases with the performance indicators obtained with other control algorithms such as CLIS [LIU 98] and Pilote (Malaterre *et al.* [MAL 95]). CLIS is based on an inverse solution method of the Saint-Venant equations, and it is designed for the automation of demand-oriented systems. Pilote is an LQR closed-loop controller and it is obtained from the steady-state solution of the Riccati equation. A Kalman filter is used to reconstruct the state variables and the unknown perturbations from a reduced number of observed variables.

10.7.1. *Test case 1-1 (Maricopa Stanfield)*

In test 1-1, GoRoSoBo shows the best values of the maximum and average MAE performance indicator, Pilote shows the second-best values of the maximum MAE and CLIS shows the second-best values of the average MAE. The maximum and average MAE values for GoRoSoBo are 8.5% and 6.9%, respectively; see Table 10.20.

	TEST CASE 1-1 TUNED-UNSCHEDULED							
	MAE (%)		IAE (%)		StE (%)		IAQ (m^3/s)	
	12-24 h		12-24 h		12-24 h		12-24 h	
	Maximum	Average	Maximum	Average	Maximum	Average	Maximum	Average
CLIS	19,9	7,8	3,8	1,8	2,6	1,4	0,2	0,1
Pilote	13,4	8,6	1,8	1,3	0,4	0,1	2,5	1,7
GoRoSoBo	8,5	6,9	1,6	1,4	2,4	1,4	8,9	5,3

Table 10.20. *The performance indicators obtained in test case 1-1 (Maricopa Stanfield)*

GoRoSoBo shows the best result of the maximum IAE indicator (1.6%) but it does not show the best result of the average IAE indicator, which is very close to the best result obtained by Pilote. The values of the average IAE indicator of both control algorithms are 1.38% and 1.28%, respectively. This is the only test in which GoRoSoBo does not show the best result of this performance indicator, and the difference with the best control algorithm (Pilote) is only 0.1%.

GoRoSoBo shows the second-best values on the maximum StE indicator for this test case (2.4%), whereas Pilote shows the best results of 0.35%. The values obtained by GoRoSoBo and CLIS of the average StE indicator are similar (1.39% and 1.4%, respectively), whereas the values obtained by Pilote are excellent (0.12%).

The GoRoSoBo algorithm gives low values of this indicator in all test cases. However, it is possible to make a modification to the control algorithm in order to reduce this indicator to zero. Once the CSI algorithm calculates, in several regulation periods, that the unscheduled offtake changes are constant and close to an exact value, CSI can fix this constant value as unscheduled offtake changes. That way, GoRoSoBo will obtain the same gate trajectories every regulation period, and so the water level at all checkpoints would be the desired water level and the canal state would be steady. The results of the StE performance indicator obtained by the GoRoSoBo algorithm are good enough in these tests to introduce this modification to the control algorithm.

On the other hand, the values of the IAQ indicator obtained by GoRoSoBo are not good. These values are far from the values obtained by CLIS, which are close to zero. We can conclude that the optimum gate trajectories obtained by GoRoSoBo show the best values of the MAE and IAE indicators, but these gate trajectories are not the best when it comes to obtaining good values of the IAQ indicator.

The IAE performance indicator establishes the total changes in the flow rate through the gates introduced by the gate trajectories in order to reach the desired final flow rate. When the unscheduled offtake changes that are introduced to the system are significant, quick changes on gate trajectories are necessary in order to recover the desired value at checkpoints as soon as possible, so that the changes in the flow rate will also be significant. It is not our primary objective to further limit the constraints of gate movements in order to obtain better values of this indicator, because then the MAE and StE indicators would be worse.

10.7.2. *Test case 1-2 (Maricopa Stanfield)*

As has happened in the previous test cases, GoRoSoBo shows the best results of the maximum and average MAE and IAE indicators. The maximum MAE value on GoRoSoBo is very similar to CLIS, 33.5% and 34.5%, respectively, although GoRoSoBo obtains excellent values of the average MAE indicator (10.3%) compared to CLIS and Pilote (14.2% and 24.9%, respectively); see Table 10.21.

	TEST CASE 1-2 TUNED-UNSCHEDULED							
	MAE (%)		IAE (%)		StE (%)		IAQ (m^3/s)	
	12-24 h		12-24 h		12-24 h		12-24 h	
	Maximum	Average	Maximum	Average	Maximum	Average	Maximum	Average
CLIS	34,5	14,2	5,0	2,0	3,6	1,1	0,2	0,1
Pilote	43,0	24,9	9,2	5,2	11,2	2,9	2,9	1,4
GoRoSoBo	33,5	10,3	5,0	1,2	2,1	0,7	6,7	3,6

Table 10.21. *The performance indicators obtained in test case 1-2 (Maricopa Stanfield)*

Taking into account the maximum IAE indicator, CLIS and GoRoSoBo obtained the best results (of around 5%) whereas Pilote obtained the worst results (9.2%). As regards the average IAE indicator, GoRoSoBo obtained the best results (1.2%), second was CLIS (2%) and third was Pilote (5.2%).

GoRoSoBo was the best in both performance indicators (StE), whereas CLIS obtained better results than Pilote. As regards the IAQ indicator, GoRoSoBo obtained values quite similar to test case 1-1; we could arrive at the same conclusion about IAQ as in the last one.

10.7.3. *Test case 2-1 (Corning canal)*

In test case 2-1, GoRoSoBo also shows the best maximum and average MAE indicator of 6.1% and 4.2%, respectively; second was CLIS, which obtained the values of 6.8% and 4.4%, respectively and third was Pilote, which obtained values of 7.5% and 4.8%, respectively.

The results about the IAE indicator are similar for all of the control algorithms; although the values obtained by CLIS for the maximum and average IAE are difficult to reduce (1.4% and 1%, respectively), GoRoSoBo obtained the best values (0.7% and 0.5%).

The maximum and average StE values are similar between the control algorithms; better results are obtained by Pilote, and GoRoSoBo shows the second best results, although the values of the StE indicator are very low in all of the control algorithms: below 0.5%.

GoRoSoBo obtains the worst values for the IAQ performance indicator in test case 2-2. Nonetheless, as we explained earlier, the main objective is to drive the state of the canal to the desired state as soon as possible; this forces the algorithm to calculate significant changes in gate trajectory in those tests, which is the opposite to moving the gates progressively in order to avoid significant changes in flow through the gates.

	MAE (%)		IAE (%)		StE (%)		IAQ (m^3/s)	
	12-24 h		12-24 h		12-24 h		12-24 h	
	Maximum	Average	Maximum	Average	Maximum	Average	Maximum	Average
CLIS	6,8	4,4	1,4	1,0	0,5	0,4	5,8	3,0
Pilote	7,5	4,8	2,0	1,3	0,2	0,1	7,6	4,7
GoRoSoBo	6,1	4,2	0,7	0,5	0,3	0,2	17,1	5,2

Table 10.22. *The performance indicators obtained in test case 2-1 (Corning canal)*

10.7.4. *Test case 2-2 (Corning canal)*

GoRoSoBo shows a maximum MAE value which is considerably lower (13.7%) than the value obtained by other controllers such as CLIS and Pilote (21.1% and 34.2%, respectively); see Table 10.23. GoRoSoBo reduces the maximum MAE indicator at 35% (CLIS) and 60% (Pilote), respectively. The average MAE indicator shows a similar tendency toward the maximum MAE: whereas GoRoSoBo obtains an average MAE value of around 7.8%, the CLIS and Pilote obtain a value of 14.9% and 17.1%, respectively.

Checking this performance indicator, GoRoSoBo obtains the better values, whereas the maximum and average IAE indicators show similar values from GoRoSoBo (6.3% and 2.1%, respectively) and CLIS (7.6% and 2.8%, respectively). Pilote obtains the worst results of this performance indicator (10.6% and 7.1%).

GoRoSoBo is not the best control algorithm of the StE indicator in test case 2-2, but the value of the maximum StE indicator is very close to CLIS: 0.8% and 0.7%, respectively. The average StE indicator for GoRoSoBo and CLIS is also very close: 0.5% and 0.4%, respectively.

	TEST CASE 2-2 TUNED-UNSCHEDULED							
	MAE (%)		IAE (%)		StE (%)		IAQ (m^3/s)	
	12-24 h		12-24 h		12-24 h		12-24 h	
	Maximum	Average	Maximum	Average	Maximum	Average	Maximum	Average
CLIS	21,1	14,9	7,6	2,8	0,7	0,4	9,7	5,5
Pilote	34,2	17,1	10,6	7,1	8,8	4,3	10,4	6,1
GoRoSoBo	13,6	7,8	6,3	2,1	0,8	0,5	15,2	11,7

Table 10.23. *The performance indicators obtained in test case 2-2 (Corning canal)*

The values of the IAQ performance indicator were not good in this test. The best values of the IAQ indicator were obtained by CLIS.

From a general perspective, the results of all of the control algorithms are similar for the performance indicators. We can conclude that GoRoSoBo is a good control algorithm. The performance indicator values obtained by GoRoSoBo are quite good, especially the MAE and IAE indicators. The biggest difference between GoRoSoBo, CLIS and Pilote is, among other things, that GoRoSoBo calculates the optimum gate trajectories using the HIM matrix and recalculates them at every time step, and so this matrix is updated in real time and it does not use approximations. In this way, the relation between the water level error and gate movement is evaluated at every time step.

We propose several changes to improve the value of IAQ in those tests. For example, we could reduce the functional constraints (dU_0max and dUmax): if the gate movements are more restricted, then there are fewer changes in gate trajectory and thus fewer changes in flow through the gates. We could also increase the duration of the operation period: if there are fewer changes in gate position during the irrigation cycle, then there are fewer changes in flow through the gates. We have to remember that all of these changes would improve the IAQ indicator, but we would obtain the worst values of the MAE and IAE indicators.

10.8. Conclusion

Once we have analyzed all of the test cases and the performance indicators, we can conclude that:

1) The test cases checked in these examples are extremely difficult, especially test cases 1-2 and 2-2. In test 2-2, a disturbance is introduced into the system with a value of 80% of the total flow rate conveyed at checkpoint 8, and the water level reaches the top of the cross-section.

2) The unscheduled deliveries calculated by CSI are quite accurate, with a maximum error of 4%. The extracted flow vector obtained by CSI is very important, because CSI obtains the current state of the canal and the disturbance. All of these data aid GoRoSoBo in calculating the optimum gate trajectories.

3) The gate constraints drive the algorithm's response when the measured water level is far from the desired level at the checkpoints, and in this way, the gate trajectory is smoother.

4) Taking into account the performance indicators, we can establish a comparison between the different control algorithms:

– The MAE is the maximum deviation of the controlled water level at the checkpoint with regard to the desired water level. From the values obtained with this performance indicator, GoRoSoBo shows the best results in all tests.

– The IAE is the integrated deviation of the controlled water level from the target water depth. From the maximum and average IAE values obtained by GoRoSoBo, this algorithm shows the best results in seven out of eight cases.

– The StE is the deviation of the controlled water level at the steady state from the target water depth, and so we only consider the last two hours of the irrigation cycle when calculating this performance indicator. GoRoSoBo obtained the best results in two cases, second best in five cases and third best in one case.

– The IAQ gives us an idea of how many changes in the flow rate are introduced by the gates in order to reach the desired flow rate at the end of the test. GoRoSoBo shows the worst values of this performance indicator. When the only objective of a control algorithm is to maintain a desired water level at the checkpoints, the gate trajectories are calculated for that objective, and some performance indicators (such as MAE/IAE) benefit more than others (such as IAQ). The IAQ indicator does not consider one important fact: the flow rate can change a lot at a target, but if the water level does not vary significantly, then the flow rate extracted by the orifice offtake will be the same because the gravity offtake is not very sensitive to small changes in the water level, and for that reason, it is vital to maintain the same water level at the target. In this sense, when significant flow changes are introduced into the system, quick changes in gate trajectory are necessary in order to recover the desired value at the checkpoints as soon as possible. For this reason, it is not our primary objective to further limit the constraints on gate movement in order to obtain better values of this performance indicator. We can conclude that the magnitude of flow changes through the gates cannot be very significant because the control algorithm maintains the measured water level close to the desired level and there is a functional constraint to restrict the movement of the gates.

5) As we explained in Chapter 8, in the case of significant unscheduled flow changes, we have to update the HIM matrix at every regulation period, because it establishes the influence of a gate movement on the hydrodynamic canal state, which is significantly modified at every regulation period during the irrigation cycle in the test cases.

6) The main objective of GoRoSoBo was achieved if we look at the results obtained in sections 10.5 and 10.7.

7) The calculation time using CSI and GoRoSoBo in a feedback strategy is quite long and it is dependent on the duration of the prediction horizon, the number of checkpoints, the number of control structures, etc. In the Corning canal, the calculation time for a prediction horizon of 12 hours is approximately 50 minutes for every regulation period, using a PC with a processor speed of 2 GHz, a RAM system of 2 GB and a hard drive of 250 Gb. We could reduce the calculation time by parallelizing the algorithm code with OpenMP, MPI or CUDA and decreasing the duration of the prediction horizon. The algorithm wastes a lot of time performing a high number of iterations in the optimization problem; when the residual value of the objective function is under 1×10^{-4}, we can finish the process. Other possibilities would be not updating the HIM when there is no significant disturbance, since this step is the most time-consuming, or using the HIM at the previous time step, whereas the other algorithm calculates the HIM at the current time step so that it can simultaneously be used by GoRoSoBo in the next time step.

Conclusion and Future Work

Summary of conclusions

The algorithms proposed (GoRoSoBo and CSI), developed and tested in this book show good results in all tests. GoRoSoBo establishes the gate trajectories that satisfy a required target, whereas CSI calculates the history of extracted flow in a canal. These algorithms can operate together in a feedback strategy during an irrigation cycle. The CSI algorithm calculates the extracted flow vector and the current canal state in every regulation time step, and the GoRoSoBo algorithm recomputes the gate trajectories, considering the extracted flow vector obtained by CSI to drive the system to the desired behavior.

Conclusions of CSI

The CSI algorithm is able to estimate any change in flow from a past time to the current time, so any disturbance that modifies the free surface is characterized by this algorithm.

The CSI algorithm uses the Levenberg–Marquardt method to solve an unconstrained optimization problem. CSI is a robust algorithm which does not need constraints and converges very quickly. The hydraulic influence matrix obtained analytically is the tool which the CSI uses to establish a relation between changes in water level at checkpoints and the extracted flow at a particular section.

The test results have shown the accuracy of this algorithm (Chapters 6 and 7). In all examples, the CSI algorithm finds the disturbances introduced into the canal. An analysis of the CSI is as follows:

– CSI is able to estimate big disturbances (20% of the total flow rate in the canal) and small disturbances (1% of the total flow rate in the canal) accurately.

– Every extraction point should have a sensor measuring the water level. But in case a disturbance is not localized at a checkpoint, it is possible to set the total water volume extracted and localize the outflow point in the canal.

– The Manning roughness coefficient is extremely important in CSI, because the algorithm is very sensitive to this coefficient. For this reason, we have to establish it accurately.

– The initial flow conditions are not important data to supply CSI, because some time after the test began, the initial conditions did not affect the results of CSI. This period depends on the canal and flow features (canal length, flow conditions, etc.).

– The water level measurements should be accurate because the results obtained by CSI are extremely dependent on the water level measures. Even an error in depth gages of 1% from the total water level is acceptable for CSI.

– The water level error must be lower than the water level variation caused by the disturbance, but CSI could not calculate the disturbance accurately.

– Once CSI calculates the extracted flow vector, the algorithm can establish the hydrodynamic canal state, water level and velocity in all sections of the canal.

Conclusions of GoRoSoBo

The GoRoSoBo algorithm is able to find the optimum gate trajectory during a predictive horizon from demand deliveries, initial gate trajectories, desired water level vector, disturbances and the current canal state which are obtained by CSI. All these data are introduced to GoRoSoBo, and the algorithm recomputes the optimum gate trajectory to maintain the desired water level constant at checkpoints.

The GoRoSoBo algorithm uses the Lagrange–Newton method to solve a constrained optimization problem. This method is considered the most efficient when you have compiled the Jacobian matrix and the Hessian matrix which are used in the computation of the gate trajectories.

The introduction of constraints was absolutely necessary to ensure stability in our optimized problem, due to inherent instability in the unconstrained problem, which is caused by the condition number of the Hessian and the HIM matrices. Not all elements of the HIM have similar values; there are gates that have a significant influence in certain checkpoints and little influence in others. This disparity of influence between elements of the matrix inevitably leads to a band matrix which is, of course, badly conditioned. This was a reason to use the Marquardt coefficient which improves the Hessian matrix condition.

The watermaster can be more or less strict about the constraints in the optimization problem, because he/she must consider the main priority of the canal. For instance, if the main priority is to maintain a constant water level at any price, the constraints on the gates would be lax, so the constraints would not have an important role in computing the gate trajectories by GoRoSoBo. Alternatively, if the main priority is to maintain the water level constant, with reduced gate movement, then the constraints would be more restrictive, so they would have an important role in the optimization problem.

As shown in Chapter 8, GoRoSoBo needs to consider a forecasted prediction of disturbances in order to calculate the optimum gate trajectories, because the algorithm has to calculate gate trajectories at the present time to drive the system in future, and some changes in gate positions at the present time are needed to prevent water level variations in the future. In the case that the disturbances disappear in the future, our forecasted prediction of disturbances would be wrong, but the results obtained with this strategy in GoRoSoBo are good. These results will be more or less accurate depending on the duration of the regulation period because CSI will calculate a new extracted flow vector in dependence of regulation period.

The performance indexes in ASCE test cases have shown the accuracy of the algorithm, especially concerning the indexes involved in the water level errors (MAE and IAE). GoRoSoBo was the best controller proposed in all test cases for these indexes. The gate trajectory obtained with GoRoSoBo is the one that gets a lower water level error at checkpoints and reaches the desired water level faster at checkpoints.

Not all results obtained by GoRoSoBo are excellent, as, for instance, the following:

– The IAQ value is much bigger in GoRoSoBo than in the other controller proposed. This index is strongly linked to the main priority of the canal. In cases where the priority is to maintain the water level constant at the checkpoints, the gate movements are significant, as are the flow rate variations through the sluice gate, so the IAQ value will also be significant.

– The GoRoSoBo algorithm obtained good results for the STE index, although the algorithm has great scope for improvement. We could restrict the gate movements when the water level error is lower than a certain threshold (we could introduce a dead band).

Future work

Following the investigations described in this book, a number of projects could be taken up as follows:

– Testing the CSI algorithm in the Canal PAC-UPC, using other pool configurations and other flow conditions. In this laboratory canal, it is possible to change the working conditions with a small amount of effort, providing an excellent opportunity to extend the results obtained in this book.

– A further study on the performance index with GoRoSoBo in other kinds of canals and in other hydrodynamic conditions used by other feedback algorithms.

– Developing a new block in our overall control diagram of an irrigation canal (Chapter 4). It would be interesting to develop the off-line parameter identification. The Manning roughness coefficient is fairly difficult to define in a canal and it can change a lot after a long period of time. On the other hand, the Manning roughness coefficient has a great weight in the Saint-Venant equation as in CSI and GoRoSoBo.

– Parallelizing the code of the GoRoSoBo algorithm with CUDA or OpenMP and speeding up the calculating process. In cases where the predictive horizon is large, the regulation interval is small, the canal length and the number of checkpoints are significant, and the calculation time of the algorithm is too long to operate in real time (and even longer if it is necessary to update the hydraulic influence matrix in every regulation time step). A real solution with the time conflict would be to parallelize the algorithm with CUDA or OpenMP to reduce the CPU time. Another possibility to speed up the calculating process would be that GoRoSoBo calculates the gate trajectories using the HIM in the previous time step, while at the same time, another algorithm calculates the HIM at the current time step which will then be used by GoRoSoBo in the next time step.

– We propose several ideas to improve the value of IAQ in these tests. We could reduce the functional constraints (dU_0max and dUmax) if the gate movements are more restricted, so would be there less changes in gate trajectories and thus less changes of flow through the gates. We could also increase the operation period, if there are less changes in gate positions during the irrigation cycle and less changes of flow through the gates. Another possibility would be to penalize the changes of flow through the gates from the objective function. The objective function only depends on the water level in our optimization problem; we could reformulate the objective function so that this function was dependent on the error in the water level at checkpoints and the flow through the gates.

Bibliography

[ABB 66] ABBOTT M.B., *An Introduction to the Method of Characteristics*, Thames and Hudson, London and American Elsevier, New York, 1966.

[ADA 09] ADAMS J.C., *The Fortran 2003 handbook*, Springer, 2009.

[AEM 09] AEMET (AGENCIA ESTATAL DE METEOROLOGÍA), BRUNET M., CASADO M.J. *et al.*, Generación de escenarios regionalizados de cambio climático para España. Ministerio de Agricultura, Pesca y Alimentación, 2009.

[AGU 12] AGUILAR J., LANGARITA P., LINARES L. *et al.*, "Adaptive predictive expert control of levels in large canals for irrigation water distribution", *International Journal of Adaptive Control and Signal Processing*, pp. 945–960, 2012.

[AKO 95] AKOUZ K., BENHAMMOU A., "Modelling and predictive control of irrigation channels", *First International Symposium on Mathematical Modelling and Simulation in Agriculture & Bio-industries*, Université Libre de Bruxelles, 1995.

[ALV 04] ALVAREZ B.X., Control predictivo de canales de riego utilizando modelos de predicción de tipo Muskingum (primer orden) y de tipo Hayami (segundo orden), Master's thesis, Technical University of Catalonia, 2004.

[ALV 13] ALVAREZ A., RIDAO M.A., RAMIREZ D.R. *et al.*, "Constrained predictive control of an irrigation canal", *Journal of Irrigation and Drainage Engineering*, vol. 139, no. 10, pp. 841–854, 2013.

[AME 66] AMEIN M., "Streamflow routing on computer by characteristics", *Water Resources Research*, vol. 2, no. 1, pp.123–130, 1966.

[AME 70] AMEIN M., FANG C.S., "Implicit flood routing in natural channels", *Journal of the Hydraulics Division*, vol. 96, no. HY12, pp. 2481–2500, 1970.

[AME 75] AMEIN M., CHU H.L., "Implicit numerical modelling of unsteady flows", *Journal of the Hydraulics Division*, vol. 101, no. HY6, pp. 93–108, 1975.

[AME 77] AMES W.F., *Numerical Methods for Partial Differential Equations*, Academic Press, Orlando, 1977.

[BAL 68] BALTZER R.A., LAI C., "Computer simulation of unsteady flow in waterways", *Journal of the Hydraulics Division*, vol. 94, no. HY4, pp.1083–117, 1968.

[BAL 85] BALOGUN O.S., Design of real-time feedback control for canal systems using linear quadratic regulator theory, PhD thesis, Department of Mechanical Engineering, University of California at Davis, 1985.

[BAL 88] BALOGUN O.S., HUBBARD M., DEVRIES J.J., "Automatic control of canal flow using linear quadratic regulator theory", *Journal of Hydraulic Engineering*, vol. 114, no. 1, pp. 75–102, 1988.

[BAU 93] BAUME J.P., SALLY H., MALATERRE P.O. *et al.*, Development and field-installation of a mathematical simulation model in support of irrigation canal management, IIMI and CEMAGREF research paper, 1993.

[BAU 97] BAUTISTA E., CLEMMENS A.J., STRELKOFF T., "Comparison of numerical procedures for gate stroking", *Journal of Irrigation and Drainage Engineering*, vol. 113, no. 2, pp.129–136, 1997.

[BAU 03] BAUTISTA A.M., STRELKOFF T.S., CLEMMENS A.J., "General characteristics of solutions to the open-channel flow, feedforward control problem", *Journal of Irrigation and Drainage Engineering*, vol. 129, no. 2, pp. 129–137, 2003.

[BAU 06] BAUTISTA E., CLEMMENS A.J., STRAND R.J., "Salt river project canal automation pilot project: simulation tests", *Journal of Irrigation and Drainage Engineering*, pp. 143–152, 2006.

[BED 02] BEDIENT B., HUBER C., *Hydrology and Floodplain Analysis*, Addison-Wesley, New York, 2002.

[BEG 05] BEGOVICH O., RUIZ V.M., GEORGES D. *et al.*, "Real-time application of a fuzzy gain scheduling control scheme to a multi-pool open irrigation canal prototype", *Journal of Intelligent Fuzzy Systems*, vol. 16, no. 3, pp. 189–199, 2005.

[BEG 07] BEGOVICH O., FELIPE J., RUIZ V., "Real-time implementation of decentralized control for an open irrigation canal prototype", *Asian Journal of Control*, vol. 9, no. 2, pp. 170–179, 2007.

[BOS 90] BOS M.G., NUGTEREN J., *On Irrigation Efficiencies*, 4th ed., International Institute for Land Reclamation and Improvement, Wageningen, 1990.

[BOS 05] BOS M.G., BURTON M.A., MOLDEN D.J., *Irrigation and Drainage Performance Assessment: Practical Guidelines*, CABI Publishing, Wallingford, 2005.

[BRO 85] BROGAN W.L., *Modern Control Theory*, Prentice Hall, 1985.

[BRU 09] BRUINSMA J., The resource outlook to 2050: by how much do land, water and crop yields need to increase by 2050?, How to Feed the World in 2050, FAO, Rome, June 2009.

[BUR 83] BURT C.M., Regulation of sloping canals by automatic downstream control, PhD thesis, Utah State University, 1983.

[BUY 79] BUYALSKI C.P., SERFOZO E.A., Electronic filter level offset (ELFLO) plus reset equipment for automatic downstream control of canals, Technical report no. REC-ERC-79-3, U.S. Department of the Interior, Bureau of Reclamation, Denver, 1979.

[BUY 91] BUYALSKI C.P., EHLER D.G., FALVEY H.T. *et al.*, Canal system automation manual, vol. 1, U.S. Department of the Interior, Bureau of Reclamation, Denver, 1991.

[CAM 04] CAMACHO E.F., BORDONS C., *Model Predictive Control*, 2nd ed., Springer-Verlag, London, 2004.

[CAR 97] CARDONA J., GÓMEZ M., RODELLAR J., "A decentralized adaptive predictive controller for irrigation canals", *Proceedings of the International Workshop on the Regulation of Irrigation Canals: State of the Art of Research and Applications*, vol. 1, pp. 215–219, Marrakech, 1997.

[CHA 90] CHAN S.P., YAO B., "Performance study of GPC for processes with time delay", *IEEE*, pp. 384–389, 1990.

[CHA 94] CHAVEZ A.A., GONZALEZ E.P., KOSUTH P. *et al.*, Informe final del proyecto automatizacion de canales, no. SH 9408, IMTA, CEMAGREF, CNA, 1994.

[CHE 91] CHEVEREAU G., Contribution à l'étude de la régulation dans les systèmes hydrauliques à surface libre, PhD thesis, Grenoble Institute of Technology, 1991.

[CHE 01] CHENTOUF B., "Robust regulation of a river reach governed by Hayami model", *Proceedings of the 40th IEEE Conference on Decision and Control*, pp. 4962–4967, 2001.

[CLE 87] CLEMMENS A.J., "Delivery system schedules and required capacities", in ZIMBELMAN D.D. (ed.), *Planning, Operation, Rehabilitation and Automation of Irrigation Water Delivery Systems*, ASCE, Portland, 1987.

[CLE 98] CLEMMENS A.J., KACEREK T.F., GRAWITZ B. *et al.*, "Test cases for canals control algorithms", *Journal of Irrigation and Drainage Engineering*, vol. 124, no. 1, pp. 23–30, 1998.

[CLE 04a] CLEMMENS A.J., SCHUURMANS J., "Simple optimal downstream feedback canal controllers: theory", *Journal of Irrigation and Drainage Engineering*, pp. 26–34, 2004.

[CLE 04b] CLEMMENS A.J., WAHLIN B.T., "Simple optimal downstream feedback canal controllers: ASCE test case results", *Journal of Irrigation and Drainage Engineering*, pp. 35–46, 2004.

[CLE 05a] CLEMMENS A.J., BAUTISTA E., WAHLIN B.T. *et al.*, "Simulation of automatic canal control systems", *Journal of Irrigation and Drainage Engineering*, pp. 324–335, 2005.

[CLE 05b] CLEMMENS A.J., STRAND R.J., BAUTISTA E., "Field testing of SacMan automated canal control system", *Proceedings of USCID Third International Conference on Irrigation and Drainage*, San Luis Obispo, 2005.

[COE 77] COEURET C., "Stabilité et précision de la régulation dynamique", *La Houille Blanche*, nos 2/3, pp. 272–277, 1977.

[COR 82] CORRIGA G., SANNA S., USAI G., "Sub-optimal level control of open-channels", *Proceedings International AMSE Conference Modelling & Simulation*, vol. 2, pp. 67–72, 1982.

[COR 83] CORRIGA G., SANNA S., USAI G., "Sub-optimal constant-volume control for open channel networks", *Applied Mathematical Modelling*, vol. 7, pp. 262–267, 1983.

[COU 62] COURANT R., HILBERT K.O., *Methods of Mathematical Physics*, vol. 2, John Wiley & Sons, 1962.

[CRA 56] CRANDALL. S.H., *Engineering Analysis: a Survey of Numerical Procedures*, McGraw-Hill, New York, 1956.

[CUN 69] CUNGE J., "On the subject of a flood propagation computation method (Muskingum method)", *Journal of Hydraulic Research*, vol. 7, no. 2, pp. 205–230, 1969.

[CUN 80] CUNGE J.A., HOLLY F.M. Jr., VERWEY A., *Practical Aspects of Computational River Hydraulics*, Pitman, London, 1980.

[DEL 86] DE LEON MOJARRO B., Contribution à l'amélioration de la gestion des périmètres irrigués, PhD thesis, USTL Montpellier, 1986.

[DEL 92a] DE LEON MOJARRO B., KOSUTH P., ARGUETA J., "Use of the SIC simulation model to improve the management of an irrigation canal in la Begoña, Guanajuato, Mexico", *CEMAGREF-IIMI International Workshop*, Montpellier, pp. 173–187, October 1992.

[DEL 92b] DELTOUR J.L., Application de l'automatique numérique à la régulation des canaux ; proposition d'une méthodologie d'étude, PhD thesis, Grenoble Institute of Technology, 1992.

[DEL 13] DELGODA D.K., SALEEM S.K., HALGAMUGE M.N., "Multiple model predictive flood control in regulated river system with uncertain inflows", *Water Resources Management*, vol. 27, no. 3, pp. 765–790, 2013.

[DOO 82] DOOGE J.C.I., STRUPCZEWSKI W.B., NAPIORKOWSKI J.J., "Hydrodynamic derivation of storage parameters of the Muskingum model", *Journal of Hydrology*, vol. 54, pp. 371–387, 1982.

[DUR 04] DURDU O.F., "Optimal control of irrigation canals using recurrent dynamic neural network (RDNN)", *Critical Transitions in Water and Environmental Resources Management*, pp. 1–12, 2004.

[ELL 70] ELLIS J., "Unsteady flow in channel of variable cross section", *Journal of the Hydraulics Division*, vol. 96, no. HY10, pp. 1927–1945, 1970.

[FAO 11a] FAO (FOOD AND AGRICULTURE ORGANIZATION OF THE UNITED NATIONS), The state of the world's land and water resources for food and agriculture: managing systems at risk, Earthscan, London, 2011.

[FAO 11b] FAO (FOOD AND AGRICULTURE ORGANIZATION OF THE UNITED NATIONS), AQUASTAT online database, Rome, available at: http://www.fao.org/nr/water/aquastat/data/query/index.html, 2011.

[FER 00] FERERES E., GOLDHAMER D.A., "Avances recientes en la programación de los riegos", *Ingeniería del Agua*, vol. 7, no. 1, 2000.

[FIL 89] FILIPOVIC V., MILOSEVIC Z., "Dyn2 method for optimal control of water flow in open channels", *Journal of Irrigation and Drainage Engineering*, vol. 115, no. 6, pp. 973–981, 1989.

[FLE 87] FLETCHER R., *Practical Methods of Optimization*, 2nd ed., John Wiley & Sons, 1987.

[FRE 85] FREAD D.L., "Channel routing", in ANDERSON M.G., BURT T.P. (eds), *Hydrological Forecasting*, John Wiley & Sons, New York, 1985.

[GAR 69] GARRISON J.M., GRANJU J.P., PRICE J.T., "Unsteady flow simulation in rivers and reservoirs", *Journal of the Hydraulics Division*, vol. 95, no. HY5, pp. 1559–1576, 1969.

[GAR 92] GARCIA A., HUBBARD M., DE VRIES J.J., "Open channel transient flow control by discrete time LQR methods", *Journal Automatica (Journal of IFAC)*, vol. 28, no. 2, pp. 255–267, 1992.

[GIL 81] GILL P.E., MURRAY W., WRIGHT M.H., *Practical Optimization*, Academic Press, 1981.

[GÓM 88] GÓMEZ M., Contribución al estudio del movimiento variable en lámina libre, en las redes de alcantarillado: aplicaciones, PhD thesis, UPC, Barcelona, 1988.

[GÓM 98] GÓMEZ M., RODELLAR J., VEA F. *et al.*, "Decentralized predictive control of multi-reach canals", *1998 IEEE International Conference on Systems, Man, and Cybernetics*, pp. 3885–3890, 1998.

[GÓM 02] GÓMEZ M., RODELLAR J., MANTECON J.A., "Predictive control method for decentralized operation of irrigation canals", *Applied Mathematical Modelling*, Elsevier, 2002.

[GOU 87] GOUSSARD J., "NEYRTEC automatic equipment for irrigation canals", in ZIMBELMAN D.D. (ed.), *Planning, Operation, Rehabilitation and Automation of Irrigation Water Delivery Systems*, ASCE, New York, 1987.

[GOU 93] GOUSSARD J., "Automation of canal irrigation systems", *International Commission on Irrigation and Drainage (ICID)*, New Delhi, 1993.

[GOU 95] GOUSSARD J., "Interaction between water delivery and irrigation scheduling", *Proceedings of the ICID/FAO Workshop on Irrigation Scheduling*, Rome, 1995.

[HEN 66] HENDERSON F.M., *Open Channel Flow*, The Macmillan Co., New York, 1966.

[HOR 13] HORVÁTH K., Model predictive control of resonance sensitive irrigation canals, PhD thesis, Polytechnic University of Catalonia, 2013.

[HUI 04] HUISING K.C., Slimme Stuwen, Master's thesis, Delft University of Technology, The Netherlands, 2004.

[ISA 11] ISAPOOR S., MONTAZAR A., VAN OVERLOOP P. *et al.*, "Designing and evaluating control systems of the Dez main canal", *Irrigation and Drainage*, pp. 70–79, 2011.

[KHA 92] KHALADI A., Gestion automatique des transferts d'eau en réseaux hydrauliques maillés à surface libre ; application au réseau des Wateringues, PhD thesis, Ecole Centrale de Lyon, CEMAGREF, 1992.

[KOS 92] KOSUTH P., ROUX A., LAFAYSSE J.L., "Use of a simulation model to improve operation of dam-river systems for irrigation purpose: elements for a methodology", *CEMAGREF-IIMI International Workshop*, pp. 81–90, 1992.

[KUH 51] KUHN H.W., TUCKER A.V., "Nonlinear programming", in NEYMAN J. (ed.), *Proceedings of Second Berkeley Symposium on Mathematical Statistics and Probability*, University of California Press, 1951.

[LEE 90] LEE T.H., LAI W.C., KWEK K.H., "Extended generalised predictive control incorporating feedforward", *Proceedings of the American Control Conference*, pp. 2703–2708, 1990.

[LEF 77] LEFEBVRE J., "La régulation dynamique, sa mise en oeuvre au canal de provence", *La Houille Blanche*, pp. 265–270, 1977.

[LEM 13] LEMOS J.M., PINTO L.F., RATO L.M. *et al.*, "Multivariable and distributed LQG control of water delivery canal", *Journal of Irrigation and Drainage Engineering*, vol. 139, no. 10, pp. 855–863, 2013.

[LIG 67] LIGGETT J.A., WOOLHISER D.A., "Difference solutions of the shallow-water equation", *Journal of the Engineering Mechanics Division*, vol. 395, no. EM2, pp. 39–71, 1967.

[LIG 75a] LIGGETT J.A., "Basic equations of unsteady flow", in MAHMOOD K., YEVJEVICH V., MILLER W.A. (eds), *Unsteady Flow in Open Channels*, Water Resources Publications, 1975.

[LIG 75b] LIGGETT J.A., CUNGE J.A., "Numerical methods of solution of the unsteady flow equations", in MAHMOOD K., YEVJEVICH V., MILLER W.A. (eds), *Unsteady Flow in Open Channels*, Water Resources Publications, 1975.

[LIN 92a] LIN Z., MANZ D.H., "Optimal operation of irrigation canal systems using nonlinear programming - dynamic simulation model", *CEMAGREF-IIMI International Workshop*, pp. 297–306, October 1992.

[LIN 92b] LINKENS D.A., MAHFOUF M., "Generalized predictive control with feedforward (GPCF) for multivariable anaesthesia", *International Journal of Control*, vol. 56, no. 5, pp. 1039–1057, 1992.

[LIN 01] LINARES P., RAMOS A., SANCHEZ P. *et al.*, Modelos Matemáticos de Optimización, Comillas Pontifical University, Madrid, 2001.

[LIR 01] LIRIA MONTAÑÉS J., *Canales hidráulicos. Proyecto, construcción, gestión y modernización*, Colegio De Ingenieros De Caminos, Canales Y Puertos, Madrid, 2001.

[LIT 03] LITRICO X., FROMION V., BAUME J.P. *et al.*, "Modelling and PI controller design for an irrigation canal", *European Control Conference*, 2003.

[LIT 06] LITRICO X., FROMION V., "Tuning of robust distant downstream PI controllers for an irrigation canal pool. I: theory", *Journal of Irrigation and Drainage Engineering*, pp. 359–368, 2006.

[LIU 92] LIU F., FEYEN J., BERLAMONT J., "Computation method for regulating unsteady flow in open channels", *Journal of Irrigation and Drainage Engineering*, vol. 118, no. 10, pp. 674–689, 1992.

[LIU 94] LIU F., FEYEN J., BERLAMONT J., "Downstream control algorithm for irrigation canals", *Journal of Irrigation and Drainage Engineering*, vol. 120, no. 3, pp. 468–483, 1994.

[LIU 95] LIU F., FEYEN J., BERLAMONT J., "Downstream control of multireach canal systems", *Journal of Irrigation and Drainage Engineering*, vol. 121, no. 2, pp. 179–190, 1995.

[LIU 98] LIU F., FEYEN J., MALATERRE P.O. *et al.*, "Development and evaluation of canal automation algorithm CLIS", *Journal of Irrigation and Drainage Engineering*, vol. 124, no. 1, pp. 40–46, 1998.

[LUE 84] LUENBERGER D.G., *Linear and Nonlinear Programming*, 2nd ed., Addison-Wesley, 1984.

[MAL 94] MALATERRE P.-O., Modélisation, analyse et commande optimale LQR d'un canal d'irrigation, PhD thesis, CEMAGREF, Montpellier, 1994.

[MAL 95] MALATERRE P.-O., "PILOTE: optimal control of irrigation canals", *First International Conference on Water Resources Engineering, Irrigation and Drainage*, San Antonio, August 1995.

[MAL 96] MALATERRE P.-O., RODELLAR J., "Multivariable predictive control of irrigation canals. Design and evaluation on a 2-pool model", *Journadas Hispano-Francesas: Sistemas inteligentes y control avanzado*, Barcelona, 1996.

[MAL 98a] MALATERRE P.-O., BAUME J.P., "Modeling and regulation of irrigation canals: existing applications and ongoing researches", *Journal of Irrigation and Drainage Engineering*, pp. 203–208, 1998.

[MAL 98b] MALATERRE P.-O., ROGERS D.C., SCHUURMANS J., "Classification of canal control algorithms", *Journal of Irrigation and Drainage Engineering*, vol. 124, no. 1, pp. 3–10, 1998.

[MAL 99] MALATERRE P.-O., BAUME J.P., SIC 3.0, a simulation model for canal automation design, CEMAGREF Montpellier, 1999.

[MAL 07] MALATERRE P.-O., CHATEAU C., "SCADA interface of the SIC software for easy real time application of advanced regulation algorithms", *Second Conference on SCADA and Related Technologies for Irrigation System Modernization – A USCID Water Management Conference*, Denver, 2007.

[MAR 89] MARZOUKI T.Z., Le système hydraulique de la rivière larcis (barrages, rivière, périmètres irrigués) : modélisation hydraulique, analyse et amélioration de la gestion, CEMAGREF, ENGREF, CARA, 1989.

[MAR 96] MARTÍN-SÁNCHEZ J.M., RODELLAR J., *Adaptive Predictive Control: from the Concepts to Plant Optimization*, Prentice Hall, 1996.

[MET 96] METCALF M., REID J., *Fortran 90, 95 Explained*, 2nd ed., Oxford University Press, 1996.

[MIN 98] MINISTERIO DE MEDIO AMBIENTE, Libro Blanco del Agua en España, Madrid, 1998.

[MIN 00] MINISTERIO DE MEDIO AMBIENTE, Libro Blanco del Agua en España, Madrid, 2000.

[MON 05] MONTAZAR A., VAN OVERLOOP P.J., BROUWER R., "Centralized controller for the Narmada Main Canal", *Irrigation and Drainage*, vol. 54, no. 1, pp. 79–89, 2005.

[MOR 93] MORÉ J.J., WRIGHT S.J., Optimization software guide, SIAM, Philadelphia, 1993.

[MOU 96] MOUSSA R., "Analytical Hayami solution for the diffusive wave flood routing problem with lateral inflow", *Journal of Hydrological Processes*, vol. 10, pp. 1209–1227, 1996.

[NAG 91] NAG, Minimizing or Maximizing a Function, The Numerical Algorithms Group Fortran Library Manual, E04, Mark 15, Oxford, 1991.

[OCA 09] OCA BARADAD J., "El regadiu en Catalunya", *Nota d'Economia*, nos 93–94, pp. 53–65, available at: http://economia.gencat.cat/web/.content/documents/articles/arxius/elregadiuacatalunya.pdf, (accessed: 05/19/2012), 2009.

[PER 89] PERRIN G., Automatisation d'un système d'irrigation à gestion centralisée, Conservatoire National des Arts et Métiers, Centre Régional Associé de Metz, 1989.

[PIQ 82] PIQUEREAU A., VILLOCEL A., Gestion automatique des eaux d'étiage ; cas de la rivière Arrats, ONERA, CERT/DERA, CACG, Toulouse, 1982.

[PLU 88] PLUSQUELLEC H., Improving the operation of canal irrigation systems: an audiovisual production, The Economic Development Institute and the Agriculture and Rural Development Department of the World Bank, 1988.

[PLU 94] PLUSQUELLEC H., BURT C., WOLTER H.W., Modern water control in irrigation, World Bank technical paper no. 246, 1994.

[PNA 11] PNACC, Segundo informe de seguimiento del plan nacional de adaptación al cambio climático, available at: http://www.mapama.gob.es/es/cambio-climatico/temas/impactos-vulnerabilidad-y-adaptacion/2%20informe%20seguimiento%20pnacc_tcm30-70404.pdf (accessed: 10/10/2013), 2011.

[PRE 97a] PRESS W.H., TEUKOLSKY S.A., VETTERLING W.T. *et al.*, *Numerical Recipes in Fortran 77: the Art of Scientific Computing*, 2nd ed., Cambridge University Press, 1997.

[PRE 97b] PRESS W.H., TEUKOLSKY S.A., VETTERLING W.T. *et al.*, *Numerical Recipes in Fortran 90: the Art of Parallel Scientific Computing*, 2nd ed., Cambridge University Press, 1997.

[QIA 10] QIAO Q.-S., YANG K.-L., "Modeling unsteady open-channel flow for controller design", *Journal of Irrigation and Drainage Engineering*, pp. 383–391, 2010.

[RED 92] REDDY J.M., DIA A., OUSSOU A., "Design of control algorithm for operation of irrigation canals", *Journal of Irrigation and Drainage Engineering*, vol. 118, no. 6, pp. 852–867, 1992.

[RED 96] REDDY J.M., "Design of global control algorithm for irrigation canals", *Journal of Hydrologic Engineering*, vol. 122, no. 9, pp. 503–511, 1996.

[ROD 93] RODELLAR J., GÓMEZ M., BONET L., "Control method for on demand operation of open-channel flow", *Journal of Irrigation and Drainage Engineering*, vol. 119, no. 2, pp. 225–241, 1993.

[ROG 98] ROGERS D.C., GOUSSARD J., "Canal control algorithms currently in use", *Journal of Irrigation and Drainage Engineering*, vol. 124, no. 1, pp. 11–15, 1998.

[ROM 88] ROMANOWICZ R.J., DOOGE J.C.I., KUNDZEWICZ Z.W., "Moments and cumulants of linearized St. Venant equation", *Advances in Water Resources*, vol. 11, pp. 92–100, 1988.

[SAB 85] SABET M.H., COE J.Q., RAMIREZ H.M. *et al.*, "Optimal operation of California aqueduct", *Journal of Water Resources Planning and Management*, vol. 111, no. 2, pp. 222–237, 1985.

[SAN 99] SANDERS B.F., KATAPODES N.D., "Control of canal flow adjoint sensitivity method", *Journal of Irrigation and Drainage Engineering*, vol. 125, no. 5, pp. 287–297, 1999.

[SAW 91] SAWADOGO S., ACHAIBOU A.K., AGUILAR-MARTIN J., "An application of adaptive predictive control to water distribution systems", IFAC, *Symposium on Intelligent Tuning and Adaptative Control*, p. 6, Singapore, 1991.

[SAW 92] SAWADOGO S., ACHAIBOU A.K., AGUILAR-MARTIN J., Long-range predictive control of an hydraulic system, CEMAGREF-IIMI workshop, 1992.

[SAW 95] SAWADOGO S., MALATERRE P.O., KOSUTH P., "Multivariable optimal control for on-demand operation of irrigation canals", *The International Journal of Systems Science*, vol. 26, no. 1, pp. 161–178, 1995.

[SCH 91] SCHUURMANS J., Models and controllers of open channel flow for irrigation purposes: a literature study, Delft University of Technology, 1991.

[SCH 92] SCHUURMANS J., Controller design for a regional downstream controlled canal, Delft University of Technology, Laboratory for Measurement and Control, 1992.

[SCH 93] SCHAALJE M., MANZ D.H., ANN controller, Personal communication, 1993.

[SEN 02] SEN D.J., GARG N.K., "Efficient algorithm for gradually varied flows in channel networks", *Journal of Irrigation and Drainage Engineering*, pp. 351–357, 2002.

[SEP 07] SEPÚLVEDA C., Instrumentation, model identification and control of an experimental irrigation canal, PhD thesis, Polytechnic University of Catalonia, 2007.

[SHA 71] SHAND M.J., Automatic downstream control systems for irrigation canals, PhD thesis, University of California, Berkeley, 1971.

[SOE 90] SOETERBOEK R., Predictive control: a unified approach, PhD thesis, Delft University of Technology, 1990.

[SOL 96] SOLER J., Integració de les equacions de Saint-Venant: aplicación al reg superficial per taules, Final project, University of Lleida, Lleida, 1996.

[SOL 03] SOLER J., Contribució a l'estudi del control de canals de regadiu per mitjà de mètodes numèrics de programació no lineal: aplicació al càlcul de les trajectòries de referencia, PhD thesis, UPC, Polytechnic University of Catalonia, 2003.

[SOL 13a] SOLER J., GÓMEZ M., RODELLAR J., "GoRoSo: feedforward control algorithm for irrigation canals based on sequential quadratic programming", *Journal of Irrigation and Drainage Engineering*, vol. 139, no. 1, pp. 41–54, 2013.

[SOL 13b] SOLER J., BONET E., GÓMEZ M., "Algoritmo 'CSI: Canal Survey Information' para el seguimiento de los caudales extraídos en canales de regadío: la protección contra los riesgos hídricos", *III Jornadas de Ingeniería del Agua*, Valencia, 2013.

[SOL 14] SOLER J., BONET E., GÓMEZ M., "CSI algorithm: Canal Survey Information to estimate the flow rate extracted from irrigation canals", *Congress on Industrial & Agricultural Canals*, Lleida, 2014.

[STR 70] STRELKOFF T., "Numerical solution of Saint-Venant equations", *Journal of the Hydraulics Division*, vol. 96, no. HY1, pp. 223–252, 1970.

[TAR 88] TARDIEU H., "Automatic dam management and river regulation for irrigation purposes", *Irrigation and Drainage Systems*, no. 2, pp. 53–61, 1988.

[THE 99] THEODORE W.M., CAIN J.T., Genetic algorithms vs. simulated annealing: a comparison of approaches for solving the circuit partitioning problem, University of Pittsburgh, available at: http://www.pitt.edu/~twmst5/tr-96-101.html, (accessed: 01/25/2011), 1999.

[TOM 89] TOMICIC B., A general optimization module for real-time control of surface water resources, Thesis, The International Institute for Hydraulic and Environmental Engineering, Delft University of Technology, 1989.

[TOU 94] TOUDEFT A., KOSUTH P., GALLINARI P., "A PID Neural Controller for Unstable Delayed Linear Systems", *Proceedings of the International Conference on Artificial Nevral Networks*, 1994.

[UND 09] UNDESA (UNITED NATIONS DEPARTMENT OF ECONOMIC AND SOCIAL AFFAIRS), World population prospects: the 2008 revision, highlights, Working paper no. ESA/P/WP.210, New York, 2009.

[VAN 05] VAN OVERLOOP P.J., VAN SCHUURMANS J., BROUWER R. *et al.*, "Multiple-model optimization of proportional integral controllers on canals", *Irrigation and Drainage Engineering*, vol. 131, pp. 190–196, 2005.

[VAN 06] VAN OVERLOOP P.J., Model predictive control on open water systems, PhD thesis, Delft University of Technology, 2006.

[VAN 08] VAN OVERLOOP P.J., WEIJS S., DIJKSTRA S., "Multiple model predictive control on a drainage canal system", *Control Engineering Practice*, vol. 16, no. 5, pp. 531–540, 2008.

[VAN 10] VAN OVERLOOP P.J., CLEMMENS A.J., STRAND R. *et al.*, "Real-time implementation of model predictive control on Maricopa-Stanfield irrigation and drainage district's WM canal", *Journal of Irrigation and Drainage Engineering*, vol. 136, no. 11, pp. 747–756, 2010.

[VIA 04] VIALA Y., MALATERRE P.-O., DELTOUR J.-L. *et al.*, "A multivariable approach for the command of Canal de Provence Aix Nord water supply subsystem", *Proceedings of Water Rights and Related Water Supply Issues USCID Conference*, Salt Lake City, 2004.

[VIE 95] VIESSMAN W., LEWIS G., *Introduction to Hydrology*, 4th ed., Harper Collins, 1995.

[WAH 03] WAHLIN B.T., BAUTISTA A.M., "Feedforward control: volume compensation versus model predictive control", *USCID-Proceedings International Conference*, Phoenix, 2003.

[WAH 04] WAHLIN B.T., "Performance of model predictive control on ASCE test canal", *Journal of Irrigation and Drainage Engineering*, pp. 227–238, 2004.

[WAH 06] WAHLIN B.T., CLEMMENS A.J., "Automatic downstream water-level feedback control of branching canal networks: simulation results", *Journal of Irrigation and Drainage Engineering*, pp. 208–219, 2006.

[WAL 87] WALKER W.R., SKOGERBOE G.V., *Surface Irrigation, Theory and Practice*, Prentice Hall, Englewood Cliffs, 1987.

[WEY 03] WEYER E., "LQ control of irrigation channel", *Proceedings of the 42nd IEEE Conference on Decision and Control*, pp. 750–755, 2003.

[WYL 69] WYLIE E.B., "Control of transient free-surface flow", *Journal of the Hydraulics Division*, pp. 347–361, 1969.

[YEV 75] YEVJEVICH V., *Unsteady Flow in Open Channels*, vol. 1, Water Resource Publications, Fort Collins, 1975.

[ZAG 92] ZAGONA E.A., Model predictive control of automated canals, PhD thesis, University of Colorado, 1992.

[ZIM 87] ZIMBELMAN D.D. (ed.), *Planning, Operation, Rehabilitation and Automation of Irrigation Water Delivery Systems*, ASCE, Portland, 1987.

Index

L, M, N

O, P, R

S, T, U, W

FSC
www.fsc.org
MIX
Papier | Fördert
gute Waldnutzung
FSC® C083411

Kolibri 360 GmbH
Ferdinand-Jühlke-Straße 7
99095 Erfurt, Deutschland
produktsicherheit@kolibri360.de